T0361395

INNOVATION IN MUSIC

Innovation in Music: Innovation Pathways brings together cutting-edge research on new innovations in the field of music production, technology, performance, and business. With contributions from a host of well-respected researchers and practitioners, this volume provides crucial coverage on the relationship between innovation and rebellion.

Including chapters on mixing desks, digital ethics, soundscapes, immersive audio, and computer-assisted music, this book is recommended reading for music industry researchers working in a range of fields, as well as professionals interested in industry innovations.

Jan-Olof Gullö is Professor in Music Production at the Royal College of Music, Stockholm, Sweden and Visiting Professor at Linnaeus University. His research interests include technical, entrepreneurial, and artistic aspects of creativity in music production.

Russ Hepworth-Sawyer is a mastering engineer with MOTTOsound, an Associate Professor at York St John University, and the managing editor of the *Perspectives on Music Production* series for Routledge.

Dave Hook is an Associate Professor in Music at Edinburgh Napier University. A rapper, poet, songwriter, and music producer, his research focuses on hip-hop, rap lyricism, identity, culture, and performance, through creative practice.

Mark Marrington is an Associate Professor in Music Production at York St John University, having previously held teaching positions at Leeds College of Music and the University of Leeds. His research interests include metal music, music technology and creativity, the contemporary classical guitar, and twentieth-century British classical music, and his recently published book, *Recording the Classical Guitar* (2021), won the 2022 ARSC Award for Excellence in Historical Recorded Sound Research (Classical Music).

Justin Paterson is Professor of Music Production at London College of Music, University of West London, UK. He has numerous research publications as author and editor. His research interests include haptics, 3-D audio, and interactive music, fields that he has investigated over a number of funded projects. He is also an active music producer and composer; his latest album (with Robert Sholl) *Les ombres du Fantôme* was released in 2023 on Metier Records.

Rob Toulson is Director of RT60 Ltd, who develop innovative music applications for mobile platforms. He was formerly Professor of Creative Industries at the University of Westminster and Director of the CoDE Research Institute at Anglia Ruskin University. Rob is an author and editor of many books and articles, including *Drum Sound and Drum Tuning*, published by Routledge in 2021.

Perspectives on Music Production
Series Editors: Russ Hepworth-Sawyer, *York St John University, UK,*
Jay Hodgson, *Western University, Ontario, Canada,* and **Mark Marrington,**
York St John University, UK

This series collects detailed and experientially informed considerations of record production from a multitude of perspectives, by authors working in a wide array of academic, creative and professional contexts. We solicit the perspectives of scholars of every disciplinary stripe, alongside recordists and recording musicians themselves, to provide a fully comprehensive analytic point-of-view on each component stage of music production. Each volume in the series thus focuses directly on a distinct stage of music production, from pre-production through recording (audio engineering), mixing, mastering, to marketing and promotions.

Reimagining Sample-based Hip Hop
Making Records within Records
Michail Exarchos

Remastering Music and Cultural Heritage
Case Studies from Iconic Original Recordings to Modern Remasters
Stephen Bruel

Innovation in Music
Adjusting Perspectives
Edited by Jan-Olof Gullö, Russ Hepworth-Sawyer, Dave Hook, Mark Marrington, Justin Paterson, and Rob Toulson

Innovation in Music
Innovation Pathways
Edited by Jan-Olof Gullö, Russ Hepworth-Sawyer, Dave Hook, Mark Marrington, Justin Paterson, and Rob Toulson

For more information about this series, please visit: www.routledge.com/Perspectives-on-Music-Production/book-series/POMP

INNOVATION IN MUSIC

Innovation Pathways

Edited by *Jan-Olof Gullö, Russ Hepworth-Sawyer, Dave Hook,*
Mark Marrington, Justin Paterson, and Rob Toulson

Routledge
Taylor & Francis Group

LONDON AND NEW YORK

Designed cover image: Rachel Bolton

First published 2025
by Routledge
4 Park Square, Milton Park, Abingdon, Oxon OX14 4RN

and by Routledge
605 Third Avenue, New York, NY 10158

Routledge is an imprint of the Taylor & Francis Group, an informa business

British Library Cataloguing-in-Publication Data
A catalogue record for this book is available from the British Library

Library of Congress Cataloging-in-Publication Data
Names: Gullö, Jan-Olof, editor. | Hepworth-Sawyer, Russ, editor. |
Hook, Dave, editor. | Marrington, Mark, editor. | Paterson, Justin, editor. | Toulson, Rob, editor.
Title: Innovation in music : innovation pathways / edited by Jan-Olof Gullö,
Russ Hepworth-Sawyer, Dave Hook, Mark Marrington, Justin Paterson, and Rob Toulson.
Description: Abingdon, Oxon New York : Routledge, 2024. |
Series: Perspectives on music production |
Includes bibliographical references and index.
Identifiers: LCCN 2024025900 (print) | LCCN 2024025901 (ebook) |
ISBN 9781032500560 (hardback) | ISBN 9781032500515 (paperback) |
ISBN 9781003396710 (ebook)
Subjects: LCSH: Music–Production and direction–Technological innovations. |
Sound recordings–Production and direction–Technological innovations. |
Composition (Music)–Collaboration. | Surround-sound systems.
Classification: LCC ML55 .I5685 2024 (print) | LCC ML55 (ebook) |
DDC 781.4–dc23/eng/20240809
LC record available at https://lccn.loc.gov/2024025900
LC ebook record available at https://lccn.loc.gov/2024025901

ISBN: 978-1-032-50056-0 (hbk)
ISBN: 978-1-032-50051-5 (pbk)
ISBN: 978-1-003-39671-0 (ebk)

DOI: 10.4324/9781003396710

Typeset in Times New Roman
by Newgen Publishing UK

CONTENTS

PREFACE

'You're not supposed to do that'. A phrase that is both a scolding and a drawing of the boundaries within which we are supposed to operate. It is the epitome of the socio-cultural boxes that we both find and put ourselves in, in order to make sense of the world. Boxes which, as soon as they are created, serve to limit or constrain. History is full of people who don't do what they were supposed to do; who take the (signal) path less travelled; who strike out into the unknown; who ask in response – 'why not?' Innovation comes from challenging existing paradigms; from saying 'what if we tried this?'; looking with fresh eyes and being willing to throw away the established rules.

The Innovation in Music Conference 2023 was held at Edinburgh Napier University, from 30 June to 2 July. The conference theme – 'You're not supposed to do that' – saw the coming together of scholars, practitioners, community activists and performers to question the current way of doing things, proposing new, ground-breaking approaches, imagined alternative perspectives, and futures outside the margins of traditional practices. This two-volume collection comprises the published proceedings from the 2023 conference. Book 1, *Innovation in Music: Adjusting Perspectives*, focuses on music, culture and society – interrogating, challenging and proposing new routes through, journeys towards, and viewpoints around the musical world. In Book 2, *Innovation in Music: Innovation Pathways*, the focus turns to reimagining, repositioning and revisiting perspectives in music production and music technology.

Volume 2 opens with Dylan Beattie's experiments in record cutting as a creative production method, as opposed to means of reproduction. His chapter, *Record-breaking: Palimpsestuous and Other Generative 'Record Cutting' Methodology Misadventures*, explores generative cutting approaches and the complex relationships formed between new and old layers of inscription. Following this, Darrell Mann's chapter, *98% Of 'You're Not Supposed To Do That' Innovation Attempts Fail: What Did The 2% Do?* examines why the majority of disruption attempts within music innovation fail and what can be learned from examining the 2% that succeed.

Chapter 3 sees Monica Esslin-Peard and Samuel D Loveless present, *I Want to Break Free: Challenging the Hegemony of Traditional Composition Through Improvisation, Performance, Collaboration and Sound Installation*. Here the authors ask the question of music educators – what can be done to adapt pedagogical practices to widen participation in composition and performance at secondary school? This is presented through a proposal for the beneficial impact of sound installation as an alterative musical experience. Whilst in Chapter 4, Vangelis Katsinas examines the architectural design and potential limitations of traditional mixer technologies in *Reinventing the Mixing Desk: A Comparative Review of the Channel Strip and the Stage Metaphor.* Further recalibration is asked of us in the following chapter from Martin K Koszolko and Kristal Spreadborough, *Whose D(Art)a is it Anyway? Repositioning Data and Digital Ethics in Remote Music Collaboration Software*, considering the ethics of Remote Music Collaboration, and creators' rights in a collaborative, distance environment.

In Chapter 6, Rotem Haguel and Justin Paterson combine meditative techniques with modular synthesis practice, presenting a framework for creative application of meditative contemplation to composition with the listener in mind. New means of perceiving and reacting to sound are also addressed in Pengcen Liu's chapter, *Research the Effect of Visual Stimuli on Auditory Perception in Music Recording and Listening.*

Andrew R Brown aims to look at DIY disruption, adding to the conversation in the democratisation of music technology with *The Impossible Box: Building a DIY Groovebox on a $10 Microcontroller.* Following this, Tore Teigland proposes a framework for soundscape construction at recording and mixing stages with *The Soundscape Cube System A Method for the Construction of a Coherent Soundscape During Recording and Mixing.* Continuing on a sound engineering pathway, the educational value of archival materials through access to multi-track recordings from the past is discussed by Paul Thompson, Toby Seay and Kirk McNally in their fascinating *Digging in the Tapes: Multitrack Archives as an Emerging Educational Resource*, exploring the EMI Music Canada Archive at the University of Calgary, Canada.

Chapters 11 and 12 both concern music mastering, with Russ Hepworth-Sawyer discussing spatial audio mastering and the significant changes that have taken place in recent years in *Gatekeeping in the Audio Mastering Industry*. This is followed by *Music Mastering and Loudness Practices Post LUFS* by Pål Eril Jensen, Tore Teigland and Claus Sohn Anderson, contrasting loudness recommendations and loudness levelling in contemporary mastering practices.

In Chapter 13, Juhani Hemmilä and Jason Woolley revisit LCR as a mixing and listening format, considering its potential in contemporary music production in their chapter, *LCR: A Valuable Multichannel Proposition for Modern Music Production?* Multi-channel audio is also the topic of Adam Parkinson and Justin Randell's chapter, *Rethinking Immersive Audio*, where the authors consider immersive audio in the fields of gaming, theatre and heritage, then present how these perspectives could be brought back into the field of sound and music.

Hussein Boon also asks the reader to consider how established concepts and practices may be applied in new ways. In their chapter, *Deliberate Practice and Unintended Consequences in Music Production as Practice and Pedagogy*, they propose the (mis)use of quantisers for pitch and timing to be used in a creative fashion, organising randomised data into new musical structures. Mads Walther-Hansen continues the theme of harnessing technology for creative application, with a timely investigation into the use of AI-based technologies for creative music practice in *Artistic Intuition and Algorithmic Prediction in Music Production.*

The final two chapters of the anthology continue to probe and push at the boundaries of technological and cultural intersections. Firstly, Charles Norton, Daniel Pratt and Justin Paterson describe *A Radiological Adventure: The Sonification of the Apocalypse*, whereby a Geiger counter is connected to a synthesiser in an experiment in mapping radiation levels to musical output. Lastly, in Chapter 18 Henrique Portovedo presents the hybridisation of the artistic and the computational in *Computer-Assisted Music as Means of Multidimensional Performance and Creation: A Post Approach to "Singularity Study 3".*

A heartfelt thank-you to all of the contributors, authors, presenters, performers and delegates for their input both to the conference and to the chapters of this book. These works comprise a spectral snapshot of the diverse, daring and distinctive counter-perspectives for change and innovation, revision, revolution and disruption in the future of music production, performance, industry, culture and society. Keep doing what you're not supposed to.

Dave Hook (Glasgow, Scotland), February 2024

1

RECORD-BREAKING

Palimpsestuous and Other Generative 'Record Cutting' Methodology Misadventures

Dylan Beattie

1.1 Introduction

It is over a century since Bahaus professor and artist László Maholy-Nagy published two essays promoting the potentiality of the phonograph as a production rather than solely a reproduction instrument. Maholy-Nagy's definition of this activity was as "productive creation" (Passuth and Maholy-Nagy, 1987, p. 289) and although his vision of directly inscribing discs by hand was never realised, the impact of statements he made in the 1920s continue to resonate with current DAW-oriented practice (i.e. a medium replacing the 'need' for recording in the acoustic domain and blurring composition, production and performance practices).

There is evidence of some disc-based composition in the years following Maholy-Nagy's essays. However, with the adoption of magnetic tape in the late 1940s onward, phonographic discs would move from being the primary recording technology – and thus also creative, experimental sonic medium – to one largely reserved for the translation of existing recorded performance from other media. Increasingly, disc recording became the pursuit of mastering engineers alone, somewhat obfuscated for, or largely ignored by, arts practitioners to undertake experimental record inscription techniques.

By returning to Maholy-Nagy's vision of the use of the record for "productive creation" purposes this chapter sets out to challenge some conventions held in relation to disc recording. In deliberately creating 'faulty' records using experimental inscription practices and with a focus on overwriting modulated grooves, the aim is that disc recording lathes and processes are promoted as musical instrument or creative actant. The affordances of generative audio inscription are presented as offering an extension to turntablism and sound art practices and overwriting records is linked to other forms of palimpsest.

1.2 Background and Related Work

1.2.1 Records as Productive Medium and Generative Inscription

Maholy-Nagy's articles, *Production – Reproduction* (1922) and *New Form in Music. Potentialities of the Phonograph* (1923), suggest the use of the phonographic disc as productive medium through the use of direct inscription methods (Passuth and Maholy-Nagy, 1987, pp. 289–292). He proposed

DOI: 10.4324/9781003396710-1

the creation of a "groove-script alphabet" or indexical composition system so that a composer's sonic vision could be directly etched without the need for interpretation by performing musicians. Katz concludes that Maholy-Nagy and other collated contemporaneous "visionaries" all "wanted more from the phonograph than it could feasibly allow" (Katz, 1999, p. 248). Given the complexities of inscription, these practitioners quickly moved on to working with other media – Maholy-Nagy switched to drawing sound in the optical soundtracks of film, for example (Patteson, 2015, p. 92).

Seemingly, the idea of the record as production medium was largely abandoned, however there are examples of disc-based composition: as in the repetition of locked grooves (single continuous rotation of a disc) on multiple records in early *Musique Concrète* practice in the 1940s (see for e.g. Schaeffer, 2013); the layering in Hindemith and Toch's *Grammophonmuzik* in 1930 (Katz, 2010, pp. 99–113); Les Paul's "new sound" in the 1940s (Paul and Cochran, 2016, p. 178); and works related to the RCA synthesiser in the 1950s (Olson and Belar, 1955). Each of these practices used a cutting machine and at least one additional playback turntable to record a new, combined, performance though without much focus on the act of inscription itself. Much like those before them, these practitioners would move to tape, given that disc recording was largely perceived as a hindrance.

Phonographic discs would come to have another productive role in their use, and 'abuse', in turntablism and sound art practices. Here existing grooves, or turntables, are modified for changes in tonearm tracking behaviour and the speed and direction of the record manipulated during playback. Practitioners augment or manipulate the surface of the medium with audible outcomes, foregrounding the sound of 'malfunction' as elements are repeated, clicks and pops featured, limitation and artefacts of process celebrated. There are numerous texts exploring the use of these media-based instruments in some depth, and some complexities can exist in disaggregating the practitioner's approach to the turntable and record. Kelly (2009, pp. 95–97) effectively taxonomises differing levels of adjustment in the "cracking" (largely surface changes) or "breaking" (generally destruction and/or reconstruction) of each. As combined instrument, the record player and record can be 'prepared' as in experimental performance practices with traditional instruments and the movements of sound artists and turntablists manipulating playback, such as Christian Marclay, constitute 'extended technique' related to the turntable.

Some practitioners of turntablism or sound art, such as Milan Knížák or Maria Chavez (2018), employ indeterminacy or generative processes in their preparation of records. The outcome of these processes means that the behaviour of the prepared record may well act as performance partner in improvisation practices. In the majority of works by these types of practitioners, the materials used are pre-existing or specially premade one-off records; the playback behaviour is subsequently modified, rather than engaging with sound inscription as a form of preparation or extended technique; or, in other words, the records used are still created in the 'usual' way by disc mastering specialists.

There is a subset of sound art, composition and turntablism practices employing experimental sound inscription (i.e. deliberate acts of inscription on disc using a transducer) in generative ways as part of the process of development or during the performance of a work. Table 1.1 shows a selection of these works, which were created prior to the practice outlined in Section 3 of this chapter. Some works presented can be categorised as process-driven and the behaviour caused by the inscription gives the record a 'voice' as actant, offering varying degrees of stochasticity in the audible outcome. In some cases, inscription methodology is used as a starter for further creative 'jumping off' in response to the behaviour or characteristics of the playback.

This type of generative inscription requires a shift in the thinking related to the recording equipment used to make records. Repurposing record lathes and related equipment from 'disc mastering' activities and foregrounding the inscription activity as productive, can involve

TABLE 1.1 A selected discography of sound art and composition involving generative inscription

Beattie, Dylan., Canon, Antoine and Hignell-Tully, Daniel	Brexshitting: The portmanteau of tautological research (2019)	Two separate field recordings are cut on the same side of a disc as concentric grooves. A 'lead in' groove (wide groove spacing) is then cut across the entire disc. The tonearm follows the lead in intersecting the two recordings and occasionally offers continuous rotations in an unpredictable way depending on the weight and anti-skate mechanism of the tonearm.
Corker, Adrian	Music for lock grooves (2019)	A number of locked grooves are cut to an 'acetate' disc. As it has been cut on a soft medium it is played repeatedly until they 'break down'.
Diepenmaat, Jeroen	Mes EP01 (2017)	An amplified knife cuts a locked groove repeatedly.
Kelly, James	Phonography (2023)	A collection of works made primarily during his (2019) practice-based thesis. Kelly undertakes experimental cutting methods to demonstrate the record lathe as musical instrument. In his 'cutovers' he inscribes drones on the top of existing grooves.
Leguay, Yann	Drift #02 (2016)	Twelve 'silent' intersecting grooves are inscribed to disc.
Shaefer, Janek	On/Off LP (2001)	Two inscriptions of the same drone track are cut on one side with one off-centre hole so it intersects with the 'regular' grooves in a different 'orbit'.
Shaefer, Janek	Skate LP (2003)	Shaefer struggled to find any mastering engineer willing to risk their equipment on non-standard practices during the creation of On/Off (2001) so for this work he modified a gramophone and created a series of 'sound scars' on disc himself. There is no spiral for the needle to track so the tonearm is free to move across the disc picking up sections of sound.
Tétreault, Martin., erikM, Dj Sniff, Rivière, Arnaud	Drift #01 (2012)	Four improvised musical performances are overlaid during the cutting process so that grooves intersect unpredictably on playback.

undertaking non-standard inscription paths or approaches, monitoring the outcome live with a tonearm and moving away from any notion of 'high quality' audio being the driving factor.

1.2.2 Palimpsests and Overwriting in other Disciplines

Inspired by some previous exploratory work, overwriting was chosen as the area of generative inscription to explore further. This topic was of particular interest given mastering engineers pay special attention to the quality of the unused surface and often maintain a near-sterile environment to avoid contamination as well as static build-up.

In textual studies, a palimpsest is useful as a reference point. In times of scarcity/expense of material, or if thinking had changed on a subject, old text would be scratched or washed from parchment (or similar substrate) and the new text overlaid. Later, through processing of this material, it was discovered that these underlayers could be retrieved and some texts once thought to be 'lost' have been rediscovered, such as the 'Archimedes palimpsest' (Easton and Noel, 2010).

Metaphorically, in literature-based disciplines, the notion of the palimpsest has been used to describe relationships between the written script and underlayers of texts not explicitly present but deemed to have an entanglement with the work. Although not the first to use the term, Dillon (2007, p. 5) provides a useful definition of 'palimpsestuous' as "the structure that one is presented with as a result of that [palimpsestic layering] process and the subsequent reappearance of the underlying script". Themes of palimpsests, erasure and rewriting in the study, and creation, of other artforms are prevalent too. Artist Richard Galpin (2017, p. 128) interrogates the term "additive subtraction" as applied to an artwork noting that this "suggests a play of differences, rather than an absence of a presence".

In other fields, such as architecture, archaeology or landscape studies, the relationship between differing layers of building, or material and substrates can also be discussed as being palimpsestic. In building new structures, existing elements become sub-layered, as the surface is augmented, intertwining the 'new' with the 'old'. Layne (2014, pp. 64–68) discusses numerous examples of environmental temporal layering in peat bogs and forests, specifically identifying disappearance and reappearance in these environments.

Overwriting can clearly be approached on a practical and conceptual level in many areas and the above examples are not an exhaustive list. One clear theme is the complex interrelation between past and present, as well as the idea that overwriting does not necessarily mean 'loss', but it may be more productive to consider it as a 'change'.

1.2.3 Historical Antecedents of Overwriting Sound

Reuse of material is commonplace in record manufacturing where faulty pressings and offcuts are returned to a pellet form, processed and placed back into the presses in the factory. In this use case, new grooves are technically formed on top of old material, though there will be no detectable sonic impact on the finished work. During the 'acoustic era' (1877–1925), if the recording 'wax' cylinder was sufficiently thick then it could be 'shaved' and used again. Wax used in the creation of master discs prior to the later use of 'acetates' could also potentially be melted and reformed. One text largely aimed at 'amateur recordists' even offers guidance on how to reuse previously blank 'acetates' using chemicals or other processes to "efface the grooves" Aldous (1944, p. 22). Regardless of the above, in general, cutting to disc is very much considered a 'write once' activity and professionally focused literature assumes the use of a new blank disc.

Beyond this practical reuse of material, Feaster (2011, pp. 182–189) identifies some erasing and overwriting of 'announcements' or dictation directions within foil and wax cylinder phonograph recordings. Under a subtitle of 'mixing', he presents numerous examples of what he terms "phonographic superimposition" (Feaster, 2011, p. 183) in experiments and private demonstrations in 1878 and 1887. These, and other, 'trick recording' approaches would become parlour entertainment around the turn of the century. Referencing a home recordist and Allis' 1905 article, *Fun with the phonograph*, some direction for overwriting in the same groove is presented. Guidance relating to the differing volume levels of the parts provided with a need to be conscious about not 'drowning out' the underlayer with the subsequent performance.

Les Paul experimented with some multiple grooves on disc using multiple tonearms (Paul and Cochran, 2016, p. 81), although overwriting is not mentioned. This was far removed from his most famous contribution to overwriting: 'sound on sound' using magnetic tape from 1949 onwards. The decades of experimentation and refinement that followed are well-documented and these overwriting activities have fundamentally shaped music creation practices. As a result, sound on sound with magnetic tape is predictable; 'old' sound and 'new' sound can be made to comfortably co-exist without too much impact from either direction.

1.3 Breaking Things

1.3.1 Technical Notes for/from Mastering Engineers

The principles of disc recording are relatively simple: a transducer (a 'cutterhead'), with a sharp stylus etches marks in a disc made of a relatively soft material as it moves slowly towards the centre of the record, making a spiralled groove; a tonearm with a pickup transducer and stylus, rides, and is excited by, the modulated groove and the signal is amplified accordingly. It is the refinement of this idea toward a 'high quality' and repeatable, stereo sound carrying format which is complex and nuanced (differences in amplitude in grooves can be as low as 0.5 microns). Inscription processes use one of two main techniques, either physical removal of material ('cutting'), or impressing ('embossing'), to inscribe a modulated groove. The umbrella term 'cutting' is used to describe either inscription approach in this chapter.

There are notable limitations inherent in the phonographic record format. One such impediment is that a degree of self-erasure (or overwriting) occurs toward the centre of the disc, with a loss of high frequency information (*BBC Recording Training Manual*, 1950, pp. 24–25). This is due to 'peripheral groove speed' where the velocity of the disc and available 'space' means that the stylus does not 'get out of the way' quickly enough (Boden, 1981, p. 30). The effect of this loss has impacted the sequencing of commercial album tracks, and some classical music records have been cut 'inside out' deliberately to present 'important' sections of a longer work with improved clarity away from the centre of the disc.

Many technical papers and patents exist and focus on getting the best frequency response and consistent behaviour during record inscription. High-end cutting heads employ feedback coils to reduce the effects of surface-related resonance and some are helium cooled to be able to reproduce high frequencies more effectively without overheating.

Numerous articles, equipment or recording manuals and broadcaster training documents were consulted in the development of the work presented in the following section. These texts offered good insight into some record 'defects' and the causes of faults which often stemmed from an 'inappropriate' setup or human error.

1.3.2 How (Not) to Break Important Things

Similar to Shaefer's experience whilst trying to find someone to cut non-standard records (see Table 1.1), a concern arose that scarce, sensitive mastering equipment should not be subjected to untested methodologies where possible. A recording rig was constructed to undertake the research (see Figure 1.1) and in the spirit of media-archaeology-as-arts-practice, combinations of different eras were hybridised to create a complex 'past-present' or 'present-past' (Huhtamo, 2009, p. 200) depending on one's perspective.

The setup comprises a 1930s Presto K8 lathe turntable and cutter arm, an Arduino-controlled stepper motor, and a cutterhead (used in two of the three examples below) made from an old hard drive, resulting in fairly 'lo-fi', monoaural sound. The cutterhead was driven by a cheap power amplifier board. A tonearm with a modern cartridge was mounted and the signal run through a phono preamplifier on a small mixer. The records used were either blank PETG plastic or had recently been discarded on a city street.

Several recording styli were used in the creation of the works: diamond (specifically developed for cutting plastic), embossing steel coned styli, and new-old stock 'Soviet' gemstone of unknown origin. Weight and angle of cutting styli were adjusted using a spring mechanism, large washers, shims and height adjustment on the cutting arm.

1.3.3 Practice Example One: 'Palimpsestuous Re-Edit'

Video available at: https://youtu.be/KfSSSjkOxxc

A 'backbeat' drum loop is cut over the top of an existing commercial record. No consideration is made for the relational tempo of rotation of the disc (e.g. 133.33 BPM at 33.33RPM, assuming

FIGURE 1.1 The recording setup.

4/4) to the new material, the tempo of material in the existing grooves, nor the groove spacing (the width between the grooves) of either part. The presentation of the work, in the tradition of a 're-edit', allows elements to be repeated, or taken out of their original sequence to highlight some of the behaviours that are caused by this generative process. Other than some processing (compression and equalisation) and reverb to even out difference between the 'new' and existing layers, the sound is directly from the tonearm pickup and no new elements are added during the construction.

Choice of program material, substrate record and inscription method all have an impact on the outcome in this process. In this example, a relatively sparse drum pattern and instrumental, orchestral music as an underlayer were chosen. In order to achieve the precision of depth of cut to match the existing grooves, a commercial cutterhead and a diamond-cutting stylus were used.

Palimpsestuous Re-edit offers a demonstration of a generative palimpsestic cutting process. This in turn provides a palimpsestuous response in which the existing grooves have a direct, audible impact on the resulting audio that is produced and would unlikely have been enacted any other way; the existing substrate shapes some of the behaviour of the new inscription and reveals itself audibly as the grooves and new content intersect.

The tonearm behaviour is impacted by this criss-crossing and intersection. New, seemingly synchronous, musical relationships are formed, despite the lack of *a priori* tempi considerations mentioned above. From a hauntological perspective, it would be possible to read this change as loss, as well as to focus on the combination of past and present, with new forms revealing themselves, as apparitions from the underlayers re-emerge. Not featured in the video example given, the sound in the room during the cutting process also reveals some of the impact of the existing material; the cutterhead resonates acoustically whilst the cutting stylus makes contact with the existing grooves on the record and performs their swan song as they are irretrievably modified by this process.

Another example of the impact of the substrate was in the unexpected behaviour of the 'chip' or 'swarf' (the material removed from the disc during the cutting process). In professional mastering setups this is usually removed using a vacuum system, although in older, portable disc recorders the engineer is expected to brush this away. When the conditions are correct (i.e. heat, depth, angle) the swarf will often wrap itself up in the centre, sliding across the disc. Given that the path to the centre was not flat, the material got stuck in the existing grooves, causing audible distortion on playback, changes to tonearm behaviour and 'balling' on the cutting stylus impacting the inscription process. As response to this, the resultant disc from the process was recorded digitally twice, once with these artefacts left in the grooves, and once when the errant swarf had been thoroughly removed.

Palimpsestuous Re-edit shows using generative inscription as productive starting point for further musical work. Given the stochastic nature of a process-driven outcome, it is possible, or even necessary depending on the goal, to selectively choose what to highlight in a composition or presentation. In the creation of this work, there was much material that was generated which was similar sounding, less 'interesting' or useful in this context. Many relationships that are formed between the parts were not aesthetically pleasing. Several iterations of the work were created and some musics certainly appear more receptive than others to this particular treatment.

1.3.4   Practice Example Two: 'We Have to Keep Digging'

Video available at: https://youtu.be/-dYoakEImaQ

Taking a somewhat combative, or anti-programmatic (see for e.g. Latour and Akrich, 1994) approach, to 'defects' or fault tables described in a number of recording manuals and literature (Aldous, 1944, p. 23; Boden, 1981, p. 39; Presto Corporation, c1941, p. 23), the sound artwork *We have to keep digging* takes overwriting a stage further by repeatedly inscribing sound on a locked groove.

'Groove wall breakdown', also described as 'cut over' (Owen and Bryson, 1931), 'crossover' or 'over cutting', was the defect of interest during the development of this work. In most of the literature surveyed, the focus was on over-modulation caused by excessive amplification with not enough space between the grooves. In this scenario, the cutting stylus moves into existing grooves, or in the path of future grooves about to be inscribed. One audible result of this can be a 'pre-ghost' or echo effect, depending on the direction of travel, and can also cause the needle to jump or track incorrectly. Another way that breakdown can occur is by creating grooves that are too wide. The manual for the Presto 6N recorder makes the relationship between the width of grooves and depth of cut clear. In warning of the dangers of cutting too deeply, it notes that "it will wear out the cutting needle quickly and ruin the quality of reproduction" (Presto Corporation, c1941, p. 14).

Starting with a blank disc, the inscription takes place using an embossing cone and the modified hard drive cutterhead with a significant amount of weight added. The tonearm tracks the groove so that each layer, and the relationship between previous layers, can be interrogated whilst the inscription takes place. Initially, the signal sent to the cutterhead comprises samples of spades digging in various materials (coal, earth, gravel, sand, etc.) randomised to appear within a single rotation of the disc. Reverb can be added to signify depth and the samples, including a voice stating the title of the work, can be triggered manually at any point. The piece is finished when the speech is no longer intelligible during repetition. Note that in the linked example, there is some audible pass-through from the cutterhead which is running at a high amplitude, to the pickup which is in contact with the surface material, and, due to the nature of the setup, quite close and affected by the electromagnetic disturbance of the voice coil.

At the start of the work new audible elements are quite clearly distinct, with each layer reasonably distinguishable upon subsequent rotation. If the new layer added is sparse in nature (such as at 4:05–4:25 in the linked video) then the effect is that the previous layer reduces in amplitude. This is similar to the behaviour of an impulse sent to a tape-based delay with a clean 'middling' feedback value, acting as a sort of echo or spectre which dissipates but is present for some time as reduces (and changes character) with each repeat. In other scenarios, where the layers are more dense, the outcome is more involuted, nuanced and less predictable. Given that the existing modulation in the groove means that the new inscription is not taking place on a flat surface, partial elements of repetition of existing, and new, audio occur.

As the groove gets deeper and more complex, the sound of the inscription itself becomes more pronounced and as the cutting stylus wears, resonance increases. It is likely that friction caused by repeated inscription and playback causes further change too. The tonearm needle sits ever lower in the groove, potentially damaging the stylus – certainly at odds with mastering practices – as it jumps around the widening shape. During the inscription process, artefacts, which could be described as 'grains' of material, rather than swarf, may have further damaged the playback stylus which in turn, could also add audible effects in the groove.

Figure 1.2 shows a microscopic image of a section of the final groove to show width change. For reference, an unmodulated groove is presented immediately to the right, and on the far edge there is also a modulated groove. Modulation is clearly present in the complex groove shape, but it is also possible to have a sense, at least visually, of why the playback stylus had difficulties in translating speech, around 20 minutes into the inscription process.

During the making of this work, in one disc recording glossary from the 1940s, a timely reminder of the differences between 'fidelity' and 'quality' – still often misused synonymously – was found. Whether the resulting sounds present in *We have to keep digging* could truly be described as a

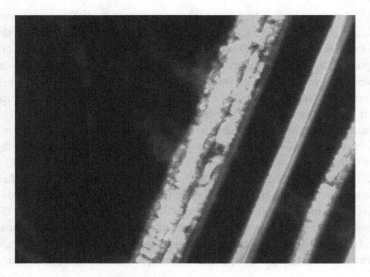

FIGURE 1.2   Microscopic image of final locked groove.

"pleasing tone" (Dorf, 1948, p. 95) is up for debate, though certainly there is very significant characteristic change evident in the outcome. Here there is an argument for neither considering the sound 'high' or 'low quality' but considering the qualities [plural] of the medium and sound produced which are affected by this overwriting intervention.

1.3.5 Practice Example Three: 'Tattoo Parlour'

Video available at: https://youtu.be/2Czd0cbQGWI

Grooves are incised by human agency into the wax plate . . . a fundamental innovation in sound production (of new, hitherto unknown sounds and tonal relations) both in composition and in musical performance.

Maholy-Nagy (1922), in Passuth (1987, p. 289)

There are improvisations on the wax plate to be considered, the phonetic results of which are theoretically unforeseeable, but which may permit us to expect significant incentives since the instrument is rather unknown to us.

Maholy-Nagy (1923), in Passuth (1987, p. 292)

A theme of precision unified all literature related to disc recording surveyed during the creation of the works presented. Instructions for avoiding faulty records, and direction towards the *correct* depth, the *correct* angle, the *correct* relationship of cutting approach to material used, all contribute to the most effective way to represent a performance. What most disc inscription practices appear to lack is much agency over these parameters during the inscription activity itself. *Tattoo Parlour*, is in part a response to this lack of agency and Maholy-Nagy's essays relating to inscription by hand.

In order to partially realise Maholy-Nagy's vision the suggestion from Magnusson (2019, p. 137) that a synthesiser would likely be part of the system was used. Partially is the operative word in this case given there are somewhat different aims: intention for musical composition, but not 'offline' inscription as Maholy-Nagy appears to suggest. The main area of consensus is the notion that the medium might allow for improvisation.

Given that the cutterhead is handheld, the performer is offered the freedom to adjust depth, to overwrite directly in the groove, inscribe on and off groove (breaking the groove wall), with differing shapes and angles of cut all causing multiple characteristic changes whilst the disc is spinning. To this end, this 'mishandling', or microvariations in handheld inscription, cause unexpected and sometimes pleasing novel audible responses for the human performer to respond to; lathe and inscription are present as actant or performance partner and being able to trigger different waveshapes and frequencies of an oscillator using a MIDI controller means that the choice of new inscription tone can be in direct response to the behaviour of the medium and inscription as a composition develops.

In the example video link, the tonearm is locked in position to allow for a performance to be somewhat pedagogic in nature, or to offer some degree of transparency. If it were free to wander, the tonearm would start tracking these grooves in unpredictable ways, given that the non-standard grooves do not necessarily follow a spiral pattern (see Figure 1.3 for example 'tattooed' record). This would of course, offer further 'voice' to the inscription methodology during improvisation.

During the creation of the work, a range of creative affordances were observed: low frequency oscillation of the cutterhead produced unexpected tones and timbre within an audible range;

FIGURE 1.3 A 'tattooed' record.

inscribing in circular patterns, or cross grooves, gave audible rhythmic outcomes; angle and depth, as in *We have to keep digging*, caused tonal characteristic change; unusually, by being able to change the angle of the cutting stylus on the horizontal axis, differences in the stereo field were created, despite the use of a monoaural cutterhead.

1.4 Conclusions

The type of generative cutting employed could be considered to extend notions of preparation of records in sound art and turntablism practices by undertaking methodologies offering new modulated grooves during the process, rather than foregrounding malfunction. Extended techniques of the record lathe and inscription are therefore being used, adding in a third component to the combination of the record and turntable as instrument in these practices. A new term for this audio augmentation would be useful as the act continues to be a surface manipulation, but one with intentional sonic layering or other manipulation effect. Feaster's "phonographic superimposition" (Feaster, 2011, p. 183) does not appear wholly suited to cover this deliberate act of 'cutting through' as well as on top of existing modulation in this way.

Much like other forms of palimpsest, the layers of the substrate contribute something significant to the outcome of the groove overwriting process; unlike other sound recording formats, the existing sound material is more likely to only be partially erased in the process, leaving remnants, a ghost in the grooves, or another tangible playback behaviour. Depending on the method of overwriting, signals are left in full to trace as an intercepted groove before returning to the new audio, offering complex relationships between time and space too. In part, this seems to fit with palimpsestic layering in other artforms where the act of overwriting itself is erasure; the surface is not erased before recording but instead whilst it is being recorded and simultaneously resonates the recording equipment. Finding a way to record this resonance successfully might offer some further insight into expected outcomes of generative overwriting processes.

In contrast with other palimpsests, overwriting in disc inscription only permits further indentation or removal of material. The title and theme of *We have to keep digging* plays on this idea, and although it might be possible to think of each level of overwriting as a 'layer', further modification is being undertaken in a way which is impossible to retrieve in a physical sense. This does not appear to distract from the overall complex themes present, as in other overwriting: loss, change, presence, and interrelations between past and present in the presented forms.

It is noted that in the three examples of practice given, the turntable movement itself was not modified as in many turntablism practices, but of course, this remains a possibility to explore in further practical experimentation.

Palimpsestic inscription is not merely a process. As in Layne's (2014, pp. 64–68) discussion of environmental palimpsests, the physical 'landscape' of the record, material, human, and technological interaction all impact the way the grooves behave and are shaped. This is particularly prevalent in *Tattoo Parlour*. Artistic and aesthetic choices are present during the development and performance of the works too. *Palimpsestuous Re-edit*, in particular, deliberately limits the sonic material being overlaid to a relatively sparse drum pattern, to allow some 'space' for the existing material to remain audible and allows for some elements of the original grooves to be relatively unchanged sonically by the process. The depth and method of the inscription, the groove spacing and the volume of the cutterhead are all also variables which have a direct impact on the outcome and could be explored further.

As in so many practices, undertaking non-standard approaches can yield valuable unexpected sonic patterns and behaviours. Working 'outside of the lines' in disc inscription during the development of the works presented has revealed several new ways of working with the medium that could be exploited in additional creative practices.

References

Aldous, D.W. (1944) *Manual of Direct Disc Recording*. London: Bernards.

BBC Recording Training Manual (1950). London: The British Broadcasting Corporation.

Boden, L. (1981) *Basic Disc Mastering*. Florida: Full Sail Recording Workshop.

Dillon, S. (2007) *The Palimpsest: Literature, Criticism, Theory*. London: Continuum (Continuum literary studies series).

Dorf, R.H. (1948) *Practical Disc Recording*. New York: Radcraft Publications.

Easton, R.L. and Noel, W. (2010) 'Infinite Possibilities: Ten Years of Study of the Archimedes Palimpsest', *Proceedings of the American Philosophical Society*, 154(1), pp. 50–76.

Feaster, P. (2011) '"A Compass of Extraordinary Range": The Forgotten Origins of Phonomanipulation', *ARSC Journal*, 42(2), p. 163.

Galpin, R. (2017) 'Erasure in Art: Destruction, Deconstruction and Palimpsest', in S. Spieker (ed.) *Destruction*. London: Cambridge, Massachusetts: Whitechapel Gallery: The MIT Press (Whitechapel: documents of contemporary art), pp. 127–131.

Huhtamo, E. (2009) 'Resurrecting the Technological Past: An Introduction to the Archeology of Media Art', in E.A. Shanken (ed.) *Art and Electronic Media*. 1st edition. London; New York: Phaidon Press, pp. 199–201.

Katz, M. (1999) *The Phonograph Effect: The Influence of Recording on Listener, Performer, Composer, 1900–1940*. Ph.D. University of Michigan. Available at: https://search.proquest.com/docview/304517668/abstract/B6048518B75B420BPQ/1 (Accessed: 29 August 2019).

Katz, M. (2010) *Capturing Sound*. First Edition, Revised ed. edition. Berkeley: University of California Press.

Kelly, C. (2009) *Cracked Media: The Sound of Malfunction*. London: MIT Press.

Latour, B. and Akrich, M. (1994) 'A Summary of a Convenient Vocabulary for the Semiotics of Human and Nonhuman Assemblies', in W.E. Bijker and J. Law (eds) *Shaping Technology / Building Society: Studies in Sociotechnical Change*. London: MIT Press, pp. 259–264.

Layne, M.K. (2014) 'The Textual Ecology of the Palimpsest: Environmental Entanglement of Present and Past', *Aisthesis. Pratiche, linguaggi e saperi dell'estetico*, 7(2), pp. 63–72. Available at: https://doi.org/10.13128/Aisthesis-15290

Magnusson, T. (2019) *Sonic Writing: Technologies of Material, Symbolic, and Signal Inscriptions*. London: Bloomsbury Academic.

New Sounds Presents: Maria Chavez (2018). Available at: www.youtube.com/watch?v=ruDZM-mrTpA (Accessed: 22 August 2023).

Olson, H.F. and Belar, H. (1955) 'Electronic Music Synthesizer', *The Journal of the Acoustical Society of America*, 27(3), pp. 595–612.

Owen, W.D. and Bryson, H.C. (1931) '*Defects in Gramophone Records*', *The Gramophone*, December, pp. 303–306.

Passuth, K. and Maholy-Nagy, L. (1987) *Moholy-Nagy*. Edited by K. Passuth. Translated by É. Grusz et al. London: Thames & Hudson.

Patteson, T. (2015) *Instruments for New Music: Sound, Technology, and Modernism*. New York: University of California Press.

Paul, L. and Cochran, M. (2016) *Les Paul: in his own words*. Centennial edition. Milwaukee, WI: Backbeat Books.

Presto Corporation (c1941) *Disc Recorder 6N*. New York: Presto Recording Corporation.

Schaeffer, P. (2013) *In Search of a Concrete Music*. trans. by Christine North and John Dack. Berkeley: University of California Press.

iscography

Beattie, Dylan., Canon, Antoine & Hignell-Tully, Daniel (2019), [4 x 7" vinyl, plus 'Union Jack' suitcase record player] *Brexshitting: The portmanteau of tautological research*, Difficult Art and Music.

Corker, Adrian (2019), [vinyl LP] *Music For Lock Grooves*, SN Variations.

Diepenmaat, Jeroen (2017), [digital release] *Mes EP1*, Self-released.

Kelly, James (2023), [digital release] *Phonography*, Self-released.

Leguay, Yann (2016), [10" vinyl] *Drift #02*, Art Kill Art.

Shaefer, Janek (2001), [vinyl LP] *On/Off LP*, AudiOh! Recordings.

Shaefer, Janek (2003), [vinyl LP] *Skate LP*, AudiOh! Recordings.

Tétreault, Martin., Dj Sniff, Rivière, Arnaud (2012), [double vinyl LP] *Drift #01*, Art Kill Art.

2

98% OF 'YOU'RE NOT SUPPOSED TO DO THAT' INNOVATION ATTEMPTS FAIL

What Did The 2% Do?

Darrell Mann

2.1 Introduction: What Is Innovation?

By way of an admittedly non-scientific experiment, a list of innovation attempts were shown to attendees at this year's IIM conference along with the question, 'how many innovations can you see?' The list comprised:

- Birotron synthesiser
- North drums
- Public Image Ltd second album, *Metal Box*
- Travis Bean aluminium-neck guitar
- MySpace Music
- Marillion album *Anoraknophobia*
- Gibson self-tuning guitar
- Amp-Lamp
- John Birch 'Magnum' guitar pickup,
- Neve 8048 mixing desk
- quadrophonic hi-fi
- Brian Hyland novelty hit-single 'Itsy Bitsy Teenie Weenie Yellow Polka Dot Bikini'
- Elvis Presley debut single 'That's All Right'
- Nick Drake third album *Pink Moon*

The methodology of the experiment might have been questionable, but the unanimity of the results perhaps conveyed a meaningful answer anyway. The consensus was that all fifteen of the examples were innovations.

Equivalent questions posed across other domains have revealed a much broader range of answers. Usually, the spectrum spans the range from zero to fifteen, typically with a bias towards the low end of the scale. The question is a difficult one because the world doesn't currently have a clear definition for what 'innovation' actually is [SIEZ, 2020]. Some might argue that the question is irrelevant. The argument made in this paper is that it matters a lot. If, for example, an author uses the definition that innovation is 'novel ideas', anyone following the advice of said author – assuming the author has been scientific during their investigation – should also expect to deliver

DOI: 10.4324/9781003396710-2

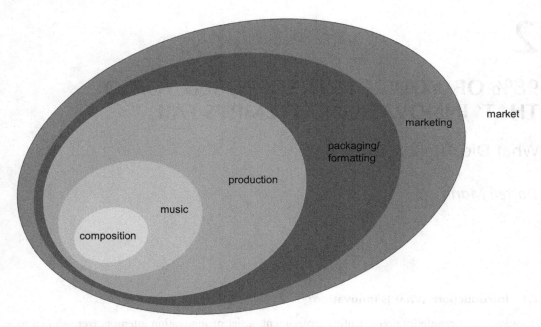

FIGURE 2.1 Hierarchical view of music industry innovation segments.

novel ideas. Similarly, if the author goes further and defines innovation as 'novel ideas that have been implemented', and is diligent in their research, followers of the 'recipe' should also expect to achieve implemented novel ideas.

The suggestion of this author is that the only useful definition of innovation is one that incorporates an appropriate measure of success: did the innovation attempt deliver a positive ROI? improve quality of life? contribute to shareholder value? Different domains will tend to use different definitions of what 'success' means to them, but ultimately, an attempt ought only to be called 'innovation' when the success criteria specified at the beginning of a project have been met. Were this criterion to be applied to the innovation candidates of Figure 2.1, the answer to the original 'how many?' question would have been three.

No individual – including this author – of course, gets to decide what 'innovation' is or is not. The importance of using a definition, however, that incorporates success is that, again assuming the requisite scientific rigour is put in place, any prospective innovator following the 'recipe' of a methodology designed around achieving 'success' should expect to achieve such a goal themselves. The question then arises, 'can there be such a thing as an innovation methodology capable of reliably and repeatedly delivering such outcomes?'

2.2 A Crash-Course Introduction To Triz

As with many major discoveries, the beginnings of TRIZ involve a happy accident. In 1946, a group of Soviet Union Navy engineers were dispatched to the Patent Office with a goal of finding 'good' patents. The immediate problem attached to this question was how to establish what 'good' means. Lead by Genrich Altshuller, the engineers determined that the only way they could meaningfully answer that question was to look at inventions that had been successfully deployed and become commercially profitable. Unlike other definitions of innovation – which predominantly either assume it to mean, as indicated in Section 1, 'novel ideas' or 'implemented novel ideas' – this one ended up eliminating close to 98% of all the patents the team examined. The happy accident here being that the decision to include 'success' as part of their innovation criteria eliminated an enormous amount of noise, and – better yet – very quickly began to reveal the signal 'DNA' of what the 2% of successful innovators did that the other 99% didn't. There were four key findings [Altshuller, 1988]:

A) Successful solutions deliver higher customer value than existing solutions, where 'value' is defined as benefits divided by the sum of cost and other 'harmful' factors.

B) Successful solutions transcend the trade-offs and compromises that most designers take for granted. In TRIZ terms, they 'solve contradictions'. This turns out to be the most 'you're not supposed to do that' finding of the original TRIZ research. The reason being that the established 'rules' of engineering design are that it is *necessary* to make trade-offs and compromises. Most education curricula call this kind of work, 'optimisation' to make it sound like a good thing. As opposed to what it actually is: the precise opposite of innovation.

C) There are very few possible strategies for solving contradictions (as opposed to the millions of ways of solving problems badly).

D) There are clear patterns of success that allow innovators to clearly see *where* future innovation opportunities will be found (importantly this is not the same as being able to 'predict the future', largely because the issue of timing, as described in a later section, is a particularly complex one to unravel).

There is considerable mythology around the number of patents analysed by Altshuller and team, but the generally held consensus is that by the late 1970s, when the research shifted to matters of psychology and pedagogy, over a million had been examined. When the Iron Curtain fell in the late 1980s, many TRIZ researchers left the former Soviet Union and began to spread news of their findings to Western countries. The success of the dissemination was somewhat patchy, but effective enough that a number of other innovation researchers – including this author – began a programme of validation of Altshuller's findings. The original research had a primarily military application focus, now the research expanded to all domains of human and natural 'innovation' efforts. Thanks to the development of advanced AI-driven search methodologies [AULIVE, 2024, Mann et al., 2015], it has been possible to increase the number of innovation case studies to, at the time of writing, over eleven and a half million examples. Including, starting around twenty years ago, an arm of the research dedicated to innovation relating to music – initially focusing on composition, and the question of what makes a piece of music evoke positive 'wow' emotions in listeners [Mann, Bradshaw, 2013]. All of the initial 'DNA' findings have now been validated and have begun to be viewed by acolytes as 'universally' applicable heuristics. This paper sees the first time these heuristics have been collectively applied to the music domain.

2.3 Innovation Dna

The emergence of AI-driven analysis tools has further enabled a more holistic analysis of a given innovation attempt, going far beyond the patent focus of the original TRIZ research. This analysis continues to confirm the validity of the original TRIZ findings, but has enabled the addition of a number of other universal success criteria to be defined. This section of the paper will detail the six 'DNA' strands most relevant to innovation attempts in an around the various parts of what might collectively be called the 'music industry' – Figure 2.1.

In order to achieve 'success', music innovators need to meet the necessary criteria of all six of these DNA elements. As such, the overall global 98% innovation attempt failure rate perhaps becomes more understandable: innovation in the context of the six strands is a particularly cruel game in that an innovation team can execute ninety-nine steps of a process perfectly, get one wrong and end up losing all their money. Very few organisations, institutions or start-up teams at this point in human history have the necessary rigour, resilience or persistence to do much better than the overall 2% success rate. The rare few that do almost inevitably have TRIZ as a part of their innovation toolkit. For the most part, these organisations have been reluctant to share the reasons underpinning their track record of success. Which is a tad frustrating for those trying to help disseminate TRIZ thinking. Perhaps the recent decision of the Chinese Government to mandate the use of TRIZ across industries in the country may begin to resolve this problem. Hopefully before Chinese companies become as globally dominant as their present trajectory would seem to indicate.

Fortunately, at this point in time, what is being disseminated is the 'classical' Soviet version of TRIZ. In which case, possibly, the evolved, more holistic perspective reported here will permit some kind of re-levelling of the innovation playing-field. Here, then, are the six 'necessary and sufficient' strands we now hypothesise are present in successful music industry innovation attempts:

2.3.1 *You're Not Supposed to Do That*

Not always, but more often than the other five strands, the rule-breaking contradiction solution defines the start of a successful innovation project. Very often, too, the unexpected, wow-inducing solution emerges from some form of crisis. It is often said that innovators should not let a good crisis go to waste, and if nothing else, one of the few benefits of being in a crisis is that it becomes clear to those involved that the current rules no longer apply and that the usual optimisation-based change strategies are no longer going to work.

Take Marillion in 1997. Their record sales are plummeting and the band is increasingly caught in a vicious cycle in which they have no money to promote their new material in order to try and then recover the sales. Touring the US was at the heart of the vicious cycle: they always lost money on the tour. When the record company declared they were no longer prepared to take the risk, American fans asked the band if they could set up their own fund as a way of getting the band over to play. In so doing, the initiative not only transformed the life of the band, pre-funding not just the tour but the next album, *Anoraknophobia*, but also accidentally invented what the world now recognises as crowd-funding. Fortunately for Marillion, they also got the other five innovation strands right too, and so are one of the three-out-of-fifteen success stories in the Figure 2.1 list.

From a TRIZ perspective, the crowd-funding solution created by the band represents an excellent illustration of Inventive Principle 13, 'The Other Way Around'. In other words, the solution, despite its apparent novelty, nevertheless fitted within the suite of universal inventive strategies uncovered during the original TRIZ research. Inventive Principle 13 was the thirteenth generic contradiction-solving strategy revealed by the patent research. The total list of strategies, based on the current eleven and a half million case study evaluations is forty [Mann, 2018, 2019] – Figure 2.2. The fortieth strategy was added to the list in the late 1970s. None of the searching since that time has revealed anything even remotely resembling a forty-first strategy. It is not possible to yet claim that the list is 'complete' – never say never – but for all practical purposes, the list is resilient enough to now say to contradiction-solvers that either individually or in combination, the solution to their problem is contained within their scope.

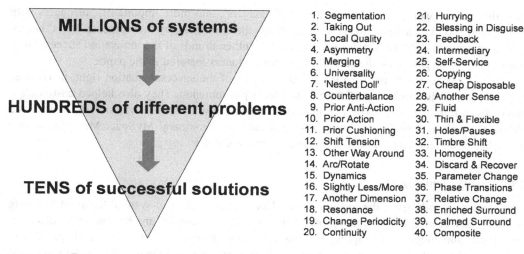

1. Segmentation	21. Hurrying
2. Taking Out	22. Blessing in Disguise
3. Local Quality	23. Feedback
4. Asymmetry	24. Intermediary
5. Merging	25. Self-Service
6. Universality	26. Copying
7. 'Nested Doll'	27. Cheap Disposable
8. Counterbalance	28. Another Sense
9. Prior Anti-Action	29. Fluid
10. Prior Action	30. Thin & Flexible
11. Prior Cushioning	31. Holes/Pauses
12. Shift Tension	32. Timbre Shift
13. Other Way Around	33. Homogeneity
14. Arc/Rotate	34. Discard & Recover
15. Dynamics	35. Parameter Change
16. Slightly Less/More	36. Phase Transitions
17. Another Dimension	37. Relative Change
18. Resonance	38. Enriched Surround
19. Change Periodicity	39. Calmed Surround
20. Continuity	40. Composite

FIGURE 2.2   Forty contradiction-resolving inventive principles.

Most of the other fifteen innovation candidates from Figure 2.1 also created step-change, 'wow'-inducing solutions by solving contradictions. The Birotron synthesiser gave musicians a step-change increase in the range of sounds that could be stored and reproduced relative to the Moog forerunner while retaining the same overall size and weight (Principles 5 and 30). The North drum offered a marked improvement in acoustic performance and simultaneously simplified manufacturability (Principle 17). The Travis Bean aluminium-necked guitar provided a much more rigid structure with a simultaneous reduction in weight that was less prone to bending as a result of the stresses created through tensioning of the strings (Principle 35). None of the three, however, make it into our Top Three places. The first two because they failed on another of the success strands. The one that scuppers the next highest proportion of all innovation attempts . . .

2.3.2   Customer Demand

An idea can be the best in the world, but if there is no market demand for that idea, it has little or no chance of being commercially successful. The usual challenge here is one of timing and the John Naisbitt [2007] quote, 'don't get so far ahead of the parade that no-one knows you're in the parade'. Sometimes the timing problem is simply that the prospective innovator is too far ahead of their time – Nick Drake for example, who was resoundingly ignored by his fellow Baby Boomers, but postmortem was adopted in far greater numbers by the much more downbeat, 'slacker', loner, Generation X [SIEZ, 2007] – but more usually the timing problem is that the innovator fails to recognise that the dynamics of a market have shifted. Both the Birotron and the North drum arrived in the mid-1970s. The 'early adopter' customers were predominantly musicians from the progressive rock genre (members of the band Yes were so enamoured of the new solutions that they invested in them [Mulryne, 2023]), a community, ironically, that was the primary target of the punk movement. By the end of 1976, the popular music industry was in the middle of a revolution. One predicated on the merits of the absence of technical prowess, and totemically, Johnny Rotten's torn 'I hate Pink Floyd' t-shirt. Musicianship was now at the bottom of the success requirement list, and having expensive flashy equipment wasn't far behind. At the other end of the timing challenge success spectrum was Elvis Presley. His contradiction-solving country-blues sound arrived at the same time as the world's first generation of 'Teenagers', and more importantly, teenagers with money to spend. And a desire to spend it on things that would annoy their parents. Elvis' launch onto the market in 1954 also managed to get all the other strands of the innovation success story together and so he gets to be the second of the three winners featured in the paper.

In theory, MySpace Music also got the timing part of the success equation right, arriving as they did right at the beginning of the Social Media phenomenon. They also helped customers – musicians in this case – to solve a contradiction around how to create awareness of your music without the assistance of the previously all-important record company. MySpace Music ticked two boxes, but sadly failed on the next of necessary success criteria . . .

2.3.3 Eco-System Buy-In

More than most industries, the music industry has existed as an 'eco-system' for almost as long as music has been a commercial proposition. As other industries are beginning to discover, one of the challenges of attempting to innovate within an eco-system is that each participant has to perceive that they will win as a result of an innovation attempt by any of the individual stakeholders. MySpace Music arrived at a time when the record companies were still getting used to the possibility that their eco-system position was under threat of being completely disrupted. In the end, record companies weren't the downfall of MySpace. Their problem was that they failed to recognise that the whole music eco-system was evolving to become a smaller part of a much bigger eco-system called Social Media [Brozyna, 2022]. MySpace thought they had found an elegant niche targeting musicians, but the bigger Social Media players quickly determined that they could serve customer needs far better by offering musicians the possibility of connecting to precisely the right audiences. Social Media platforms like Facebook and Twitter rapidly realised that whoever owns the data wins, and so made sure they owned as much data as possible. And, by literally connecting all of the Social dots, they had access to orders of magnitude more data than MySpace would ever be able to muster.

Quadrophonic hi-fi solutions fell into another variant of the eco-system trap. It is one thing to invent quadrophonic audio solutions, it is quite another to convince the recording studios to start recording quadrophonically, the record companies to press quadrophonic records, consumers to potentially have to go out and re-purchase new versions of music they already own, and, most challenging of all, to convince other hi-fi manufacturers to also transition to offering quadrophonic systems to their customers. Markets fundamentally like competition and if there is only one offeror of a given solution type, that makes anyone having to rely on the resulting monopoly reluctant to enter the game. Either everyone plays or no-one plays. And so, when it comes to quadrophonic hi-fi, despite the fact those few customers that have experienced it affirm its aural superiority, it is still little more than an obscure historical curiosity.

Gibson to some extent avoided the eco-system challenge – or at least thought they had – when they launched their self-tuning guitar solution onto the market in 2007. In theory, the innovation attempt had no impact on other parts of the industry. In practice, it at least suggested the possibility that large numbers of 'guitar techs' would no longer be required, and more often than not, in Innovation World, perception is reality. Gibson didn't get this part of the success DNA story right, but their bigger failure involved the fourth of the six essential strands . . .

2.3.4 Dynamic Measurement

Almost all organisations on the planet now understand the importance of listening to the 'Voice of the Customer'. Relatively few, however, have understood how difficult it is to capture that voice.

Especially when it comes to anything related to innovation. The first challenge here relates to the oft-used Henry Ford quotation, 'if I'd listened to the customer, they would have asked for a faster horse'. A customer that has never experienced a car before, in other words, has no words to ask for one. Smart innovators – like the second Steve Jobs era Apple – learned to tap into the unspoken needs of customers without having to directly go and talk to them [SIEZ, 2011]. What this way of thinking reveals is that more often than not, when asking customers whether they would be interested in having, say, a self-tuning guitar, what the customer says in response and the way they will subsequently behave turn out to be two very different things. This mis-communication problem then becomes exacerbated when other elements of an eco-system become engaged in the conversation. Gibson made two mistakes here. The first is misunderstanding the messages from the initial target customers – namely the novice guitarist that wants to sound proficient without the hassle normally associated with keeping a guitar in tune – about how much they would actually be willing to pay for the privilege of such convenience. The second, bigger problem, was a failure to recognise the influence that the very vocal rejection of the self-tuning solution from professional guitarists (largely due to their perception that self-tuning guitars removed some of their mystique) have on the novice guitar player. If my hero tells me something is bad, I tend to believe it is bad. Even if the reality is that it is good. 'Dynamic Measurement' in this innovation context comes down to being able to read between the lines, measure what's important rather than merely expedient, and remember that conversations – especially in a Social Media-dominated world – evolve at critical times in time-frames measurable in hours and days rather than the multi-year duration of many innovation projects. What might have resonated with potential customers, may provoke the opposite reaction tomorrow.

2.3.5   Coherent Solution

The self-tuning guitar also failed at the fifth of the six success strands. On one level, the solution offered users the contradiction-solving 'wow' of a no-effort always-in-tune guitar, but on other levels, the Company apparently got so focused on the end result, they failed to adequately address many of the new problems that their first prototypes revealed. Firstly, they didn't capture a clear understanding of how much customers would be prepared to pay for the solution and, worse, mistakenly assumed that price wouldn't be a big issue. Consequently, they didn't put enough effort into ensuring a solution had the potential to be manufactured economically. They also didn't do enough to reduce the size and weight of the device. The end result being that not only was it 'too expensive' but also, hung as it was to the back of the headstock of the guitar, it alters the balance of the instrument and hence adversely affects playability.

Lack of solution coherence – not thinking about all of the factors beyond performance: cost, reliability, maintainability, durability, sustainability, etc. – was also ultimately the downfall of Neve mixing desks. Beautiful, revered things that they were, any studio that acquired a Neve desk was fundamentally expected to take on a swathe of additional costs to keep their desk working [Taylor, 2022]. Studios want to make records not spend half their waking hours tending to temperamental and sensitive hardware.

One of the simplest, most effective niche ways to achieve 'coherence' is to create solutions with deliberately cheap, disposable ambitions. This is definitely a small niche ultimately, but one that in this case, provides the third of the three success stories from the Figure 2.1 candidate list. Namely Brian Hyland's 1960 novelty hit – the first of the 'teenager-age'. Hyland never had another Number One hit after 'Itsy Bitsy . . .' and in some ways broke the bubblegum-pop rule of having the grace to disappear from the scene after they'd committed their crime against good taste and had their success. Nevertheless, he established a formula that has subsequently allowed a steady stream of novelty acts to sell many more millions of records. The one-off novelty specialists in some ways managed to avoid having to worry about the last of the six strands, another of those that sits behind a sizeable proportion of innovation attempt failures . . .

2.3.6 Hero's Journey Persistence

Every successful innovation involves a step-change, and every step-change involves a Hero's Journey. A Journey of the sort revealed by Jospeh Campbell following a life devoted to studying the DNA of successful literature [2008]. Humans are storytellers, and most of our learning comes from stories. Campbell realised that at the heart of every successful story was a hero facing an impossible Ordeal ('contradiction' in TRIZ language) and somehow prevailed. David beating Goliath. Napster beating the Record Company establishment. The Beatles 'invading' America. But then he also realised – as illustrated in Figure 2.3 – that the Ordeal is but one stage of a much longer Journey.

Society in many ways operates in such a way as to make life difficult for innovators. As if to say to them, 'prove to me that you're not a crank like all the others; prove to me that you want this enough'. To innovate is to suffer. Or at least persist long enough through the suffering that people begin to believe that you want success badly enough. Make the innovator's life too easy

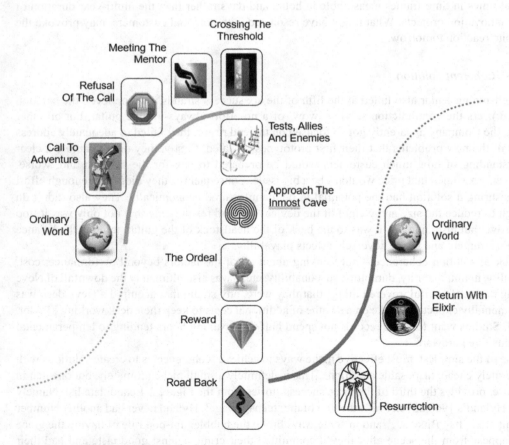

FIGURE 2.3 The hero's (innovation) journey.

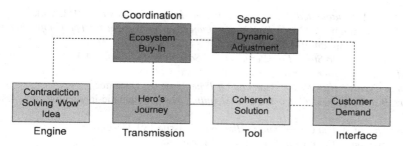

FIGURE 2.4   Minimum viable music innovation system.

and the end result is no innovation. Struggle is the soil of success. The other five DNA strands aside, if the innovator doesn't possess the requisite persistence to continue despite the oftentimes overwhelming resistance they encounter, they will end up in the 98% of other failed innovation attempts. Innovators need to 'pay their dues'. Like playing four shifts a night to drunk punters in sweaty Hamburg nightclubs for two years. No pain, no gain.

2.4   Conclusion

The paper has outlined six 'necessary and sufficient' strands that together enable innovation in the music industry. As it turns out, it is the same six in all other domains. And, further, thanks to the TRIZ research, we can say with a high degree of certainty that six is the right number. Not seven, or five. The TRIZ research also unknowingly decoded a unique first-principle understanding of 'systems'. A system in TRIZ terms is a collection of 'things' that, together deliver a useful outcome. Or, put the other way around, if there is a desire to create a useful outcome – like 'to innovate' – it will only happen if the requisite six system elements are present.

Figure 2.4 illustrates the so-called TRIZ 'Law of System Completeness' [SIEZ, 2017] incorporating the six specific elements described in Section 3 of this paper alongside the generic element labels as revealed from the TRIZ research. Think of any useful outcome that you are currently obtaining in your life – from cleaning your teeth to composing a symphony, from tuning a guitar to trying to get the three discs out of PiL's iconic-but-stupid Metal Box without scratching them – and think about the six elements and you will find that they are all present and correct. Or, put it the other way around again, think about any innovative project you're either still involved with or saw fail in the past, and you will find it didn't work because at least one of the six elements was missing completely or inadequately present. To end with a paraphrase of a favourite quote from Leo Tolstoy, 'all happy innovation projects are alike; each unhappy project is unhappy in its own way'.

References

Altshuller, G. (1988) *Creativity as an Exact Science*, Translated by Anthony Williams, Gordon & Breach, New York.

AULIVE (2024) Patent Inspiration: Creative Helicopter View.www.patentinspiration.com/, last accessed: 2024/09/27.

Brozyna, E. (2022) *What Happened to Myspace (and Is It Even Still Around)?* PureWow, 24 August.

Campbell, J. (2008) *The Hero with a Thousand Faces*, Third Edition, Pantheon.

Mann, D.L. (2018) *Business Matrix 3.0: Solving Management, People & Process Contradictions*, IFR Press.

Mann, D.L. (2019) *Oblique StrateTRIZ: Sparking Compositional Breakthrough*, IIM, London.

Mann, D.L., Bradshaw, C. (2013) *Automated 'Wow' Generation In Musical Composition*, IIM, York.

Mann, D.L., Gimenes, M., Schramm, R. (2015) *PanGenics: Objective Assessment Of The Chops, Feel & Creativity Capabilities Of A Musician*, IIM, Cambridge.

Mulryne, K. (2023) *Yes – The Tormato Story, Five Per Cent For Something Publishing*, UK.

Naisbitt, J. (2007) *Mind Set!: Reset Your Thinking and See the Future*, Collins.

SIEZ (2007) *Generational Cycles – Nick Drake: Man Out Of Time*, Issue 69, December.

SIEZ (2011) *Customers Buy Outcomes: Outcomes Buy Meaning*, www.systematic-innovation.com, Issue 110, May.

SIEZ (2017) *Case Study: Law of System Completeness Hierarchies*, Issue 189, December.

SIEZ (2020) *Defining Innovation (40 Years Too Late)*, Issue 221, August.

Taylor, D. (2022) *Early RUPERT NEVE consoles and their stories | PART TWO: 1962–1968 | 'A Revolution Has Occurred'*, postfade.co.uk.

Discography

Drake, Nick (1972), [vinyl LP] *Pink Moon*, London: Island Records.

Hyland, Brian (1960), [45rpm single] *Itsy Bitsy Teenie Weenie Yellow Polka Dot Bikini*, US: Leader/Kapp.

Marillion (2001), [CD], *Anoraknophobia*, London, UK: Liberty.

Presley, Elvis (1954), [45rpm single] *That's All Right*, Memphis, TN: Sun.

Public Image Ltd (1979), [vinyl LP] *Metal Box*, London: Virgin.

3

I WANT TO BREAK FREE

Challenging the Hegemony of Traditional Composition Through Improvisation, Performance, Collaboration and Sound Installation

Monica Esslin-Peard and Samuel D Loveless

3.1 Introduction

Music education in England and Wales would be, it seems, under threat as Cooper (2019, p.4) states 'in crisis'. Bath et al., (2020) report that music was not taught in 50% of secondary schools in the years 2017 and 2018. This should be regarded as a cause for concern, as the National Plan for Music Education (2022) states that all children should be able to play a musical instrument, sing and make music with others. The wider remit of the plan is to set out the government's vision for music education and how this can be achieved through partnerships with schools, music hubs and the music industry. In the area of West London, where this research project is based, four secondary schools do not offer music at all, despite the fact that the National Curriculum (2013) requires every state secondary school to offer music tuition to students at Key Stage 3, aged 11–14. The benefits of music education are well-recognised, as Bath et al., (2020, p.454) argue: 'The value of music in the curriculum is wide-reaching and to access it for free in school should be the right of every child'. However, government data present a very different picture. Student numbers for GCSE music, an external examination taken at age 16, and A Level music, taken at age 18, are in dramatic decline. This is partly due to government policy (DfE, 2018; Jeffreyes, 2018) introducing the EBacc, which prioritises qualifications in Maths, English, Science and Humanities over specific creative arts subjects like Art, Music and Drama, but also influenced by economic deprivation. Whittaker et al., (2019, p.1) found that lower numbers of entries of students to study A Level music in the final two years of secondary school correlate with higher levels of deprivation. It should be noted that composition at GCSE and A Level is mainly predicated upon traditional Western Classical Art music forms, such as a Bach chorale, figured bass, or theme and variations. This requires students to be immersed in classical music and be fluent in music notation and have a thorough understanding of music theory to at least ABRSM Grade 5 or, perhaps better, Grade 8 theory. Access to such specialised tuition is declining in state schools and may disappear by 2033, according to Reilly (2021). Furthermore, running A Level music with two or three music students is often deemed financially unviable by school leaders as Whittaker (2021, p.147) explains.

Government policy and economic deprivation are not the only factors to be considered. Numbers of classroom music teachers have reduced between 2016 and 2021 as Savage (2018, p.115) reports. Even more alarmingly, it would seem that the very teachers who should be inspiring students to study music in the higher years of secondary education in order to progress to the study of

DOI: 10.4324/9781003396710-3

FIGURE 3.1   QR code for audio version.

music in higher education are themselves blocking progression, as Gouzouasis and Bakan (2011, p.13) argue:

> Our profession requires a cataclysmic shift from dictating curricula and curriculum content from the 1950s – meaningless to the majority of youth for at least the past 40 years. Our teaching institutions isolate themselves from the realities of 21st century music that are founded upon numerous forms of popular music that have been the core of youth music making and music listening for at least the past 80 years. The ongoing endemic apotheosis of Western classical music and "traditional school music" must cease immediately if music educators are to remain relevant.

This view is echoed by Folkestad (2022) who advocates a mixed-media approach to composition and music making led by music technology. There have been some significant initiatives to foster music in secondary schools like Sing Up (Welch et al., 2011), Musical Futures (Hallam et al., 2008), First Access in the London Borough of Hackney (2023) and others, such as programmes based on El Sistema, (Bolden et al., 2021).

This situation is worrying, not just for secondary schools, but also for music departments in higher education. Thus, this paper considers whether adopting new approaches to music composition, improvisation and performance might increase student participation at KS3 and thus generate higher numbers of students following a pathway at KS4 (aged 14–16) and KS5 (aged 16–18) in order to study music, performance, composition, music technology or the music industry at university. The pilot study described below is founded on principles of experimental music, sound art and sound installation which may afford access to musicking to a wider range of students.

3.2   Literature Review Traditional Approaches To Composition

Traditional notation-based composition may still be regarded as the worthiest form of teaching composition as Berkley (2001, p.129) reports. There is a plethora of 'old-fashioned' composition workbooks such as Kitson (1920), Lovelock (1939, 1949), Steinitz (1967), Butterworth (1999) and many others including Fux's 'Gradus ad Parnassum' translated by Mann (1971). The challenge of this traditional and admittedly old-fashioned approach is that students not only need to be fluent in notation, but also have a wide grasp of musical styles from the Renaissance to the present day. As Durrant and Welch (1995, p.19) point out,

> Music is aural, not written: primarily to do with sound, not written symbol. The value of creative exploration of sound, as with the exploration of colour or words, lies in the creative exploration itself: the making through experiment.

The exploration of sound does not of necessity have to refer to the written symbols, traditional Western notation or any other system.

In the early 2000s, North American classroom music teachers like Beckstead (2001) were bemoaning the difficulty of teaching traditional composition in US public schools and turned towards technology to enable students to compose. However, as Winters (2012) points out, 21st-century music educators need to adopt a mixed methods approach to inspire students to compose through performance, improvisation and the use of music technology.

3.3   Considering Composition As Performance: Experimental Music?

Experimental approaches to music first emerged as a means of presenting music as performance in order to generate social meaning. This implies a move away from the text-based orientation of traditional musicology and theory which hampers thinking about music as a performance art as Cook (2001) explains. A comparable approach can be found in free ensembles, (Barber-Kersovan and Kirchberg, 2021) and Macdonald and Wilson's studies of free improvisation (2020).

3.4   What Is Sound Art Or A Sound Installation?

According to Grant et al., (2021, p.xi):

> Sound art has been resistant to its own definition. Emerging from a space between multiple movements of thought and artistic practice in the twentieth century, sound art is often discussed not in its own terms but in relation to its proximity to the music and the visual art from which it seeks to differentiate itself.

This is evident from the lack of consensus in the literature between sound art, sound installation, sonic art, sounding art and so forth. For example, a definition of sounding art posited by Cobussen et al., (2017 p.2) states: 'Human made artistic and/or aesthetic application of sounds [. . .] sounding art has the potential to teach us about what sound is or can be, how we deal with sound, what sounds can do and what it might express.' Maybe it is more useful to go back in time and consider what Varese (Varese and Wen-Chung 1966, p.16) understood by musical forms: 'The form of the work is the consequence of this interaction'. On the other hand, Strachan (2017, p.88) simply states 'Sound is just a part of everyday life. It is there. It doesn't require a musical structure'.

But perhaps this is also a matter of how we perceive sound or music, whether as a performer or an audience member, as Gibson (1979, p.63) explains: 'An individual's perception of the environment is resultant in particular courses of action whereby . . . affordances indicate the range of possible activities of a given object'.

For the purposes of this study, the term 'sound installation' is used to describe a performative act which may encompass physical movement, creating sounds with objects and creating sounds with musical instruments, including music technology, focussing on the active involvement of participants, irrespective of their musical backgrounds or abilities. Kanellopoulos (2009) assesses attempts to introduce experimental music in the classroom and concludes that the major barrier to such approaches is teachers' lack of understanding of the underlying principles of teaching works by Christian Wolff. As Nyman (1974) proposes, experimental composition is about a process of generating sounds and actions. More recently, Savage and Challis (2011, p.5) advocate composition in schools based upon 'More soundscape-oriented activity [. . .] using computer and recording technology, we are able not just to listen and appraise the sounds around us, but to sculpt with this sonic material'.

In the last ten years, researchers have explored experimental music in the classroom in the US. Some key ideas emerge here. Firstly, the classroom becomes a scene or space for collaboration (Thomson, 2007). Secondly, the teacher becomes a guide or mentor, enabling students to learn through exploring sound (Sordahl, 2013; Tinkle, 2015). Two main pedagogical approaches emerge: making sounds or creating a kind of 'noise' (Woods, 2022) and free improvisation, (Niknafs, 2013; Hickey et al., 2016).

3.5 Composition and Improvisation

How can experimental music offer student-centred learning in the field of composition? Larsson and Georgii-Hemming (2019) review 20 papers concerning pedagogical approaches to improvisation and conclude that improvisation – whether teacher led, or student led – is poorly understood in music education. Sawyer (2008), on the other hand, argues that improvisation leads to deep musical learning as improvisation involves an understanding of structure, melody, harmony, adaptation and skill in collaborating with others. Winters (2012, p.22) takes a more general view of composition, noting that 'Exploration and rehearsal of ideas, sounds and samples is the starting point for musical activity and the development of small compositions'. Such work does not need to be notation based. Nevertheless, she also recognises that many classroom music teachers find teaching composition through experimentation and improvisation challenging.

3.6 Active Listening and Collaboration In Ensembles

Such experimental approaches to composition in groups require critical listening skills in order for students to understand what works, to try out their ideas and make critical judgements about how a composition develops, as Cuervo (2018) states. As Esslin-Peard (2024) notes, learning to listen and analyse rehearsal and performance is a core expectation of classically-trained musicians, which requires sensitivities to pitch, intonation, rhythm, melody and underlying harmonic structure. Similarly, Bell (2018) advocates that musicians should focus on listening to sound, particularly in choral rehearsals. Increasingly, critical listening is recognised as a key skill for those working in production studios, (Corey, 2017). Indeed, listening is crucial for participants in a sound installation, as this study demonstrates.

3.7 Methodology

This pilot study adopts a mixed methods approach, combining quantitative and qualitative data from participant feedback questionnaires with an analysis of video recordings of five sound installation workshops. Informed written consent following BERA guidelines (2018) was obtained for all participants. Parental consent was given for primary school pupils. Everyone consented to pictures, audio and video recordings to be made during the workshops. Permission was given for this content to be used in this study.

All the workshops were facilitated by Loveless, with Esslin-Peard assisting with the school-based and amateur choral society groups. Loveless reviewed all the videos to assess to what extent he had influenced outcomes. With adults, he was able to act as a mentor/co-creator; with the Brownies, there were elements of didactic interventions when each element of the workshop was introduced. All participants took part in four works: '#1 Shuffle' by Alison Knowles (1961), an event score rooted in the work of Fluxus artists, and three compositions by Samuel D Loveless: 'The Joy of Letting Go', 'Somewhere' and 'bloom'. See Appendices 1–4 below for descriptions of each event.

3.8 Participants

A total of 85 adults and young people participated in the workshops. In the first workshop with classroom music teachers, there were 13 participants, but only six feedback questionnaires were collected, due to some participants leaving early. For the other workshops, all participants submitted questionnaires. Figure 3.2 below shows the number of participants in each location and the iterations of each element of the workshop.

For the purposes of this paper, participants are divided into adults and young people/children. The adult group includes professionally trained musicians, classroom music teachers with different levels of music training and amateur musician singers or adults working with others in musical contexts, such as the leaders supporting the Brownie group, a voluntary organisation for girls aged 7–10 years. The young people include 15–18-year-old self-declared musicians and 13–15-year-old school students who were studying music. The Brownies were not necessarily self-reported musicians, but all demonstrated some musical ability. Despite the disparate nature of the participants overall, the inclusive nature of our definition of a sound

Date	Group	Name of work/piece			
		#1 Shuffle	The Joy of Letting Go	Somewhere	bloom
		Frequency of performances			
28.2.2023	Classroom music teachers, PGCE music student N=13 Sixth form students N=3	1	1	1	2
7.4.2023	HE students and early career creatives in Sweden N=7	2	1	1	1
1.5.2023	Amateur choral society members N=16	1	1	4	1
4.5.2023	Teenage students from state secondary schools N=28 Staff N=2	1	1	1 x full ensemble 2 x small groups	2
16.5.2023	Primary school students, members of Brownies N=14 Staff N=2	2	1	2	2 x full ensemble 1 x small groups
Total	N=85				

FIGURE 3.2   Overview of workshops.

installation allows a comparison of responses to the workshops. Thus, the following research questions are posited:

How do musicians participate in four different scenarios in a sound installation?
What implications does an approach based on sound installation have for pedagogy?
How does an inclusive, participatory workshop foster skills development in composition, improvisation and performance?

3.9   Primary Research Findings

The principal research question asked was: What were the most useful aspects of today's workshop for you?

The following answers were suggested as possible responses (one or more) for the participants to choose:

Experimenting with sounds.
Using a wide range of instruments/voices/objects.
Freedom to create new sounds.
Working with participants you know well.
Working with participants you don't know well.
Listening in a different way.
Understanding new directions in sound art/sound installation.
Any other comments?

A total of 78 questionnaires was submitted, 33 by adults, 31 by older students (aged 13–18) and 14 by younger students (aged 7–10). The responses to the research question are summarised in Figures 3.3 and 3.4 below.

There was a wide range of responses. The most frequently cited responses were 'experimenting with sounds', 'freedom to create new sounds' and 'listening in a different way'. The data suggest that participants valued the opportunity to experiment. Furthermore, it would appear that classically trained musicians largely reacted to auditory and musical stimuli from those around them without feeling constrained by the environment or what they were asked to do.

The overall responses to the idea of a sound installation in Figure 3.4 highlight a comparison of the responses by age group. The data suggest that the majority of participants were positive about their experiences. The two most positive responses to the questionnaire reported by the participants were 'Experimenting with sounds' and 'Freedom to create new sounds'. Whilst adults and the primary students (Brownies) were very positive, the secondary students were less enthusiastic. This may be due to the circumstances of the workshop. Students were brought together from two single-sex secondary schools. In these circumstances, participants perhaps felt more nervous or apprehensive about fully engaging as they were in front of strangers, particularly when they were asked to shuffle. Nevertheless, the student participants greatly valued the opportunity to try out a wide range of instruments – about 400 were available to them in the school assembly hall. In the following section, highlights from the workshops with video clips are discussed.

3.10   Workshop Findings

The opening work was '#1 Shuffle' (Knowles, 1961), see Appendix 1. Examination of the video recordings shows that the youngest participants, the Brownies, and the participants in Sweden were the least inhibited in their responses.

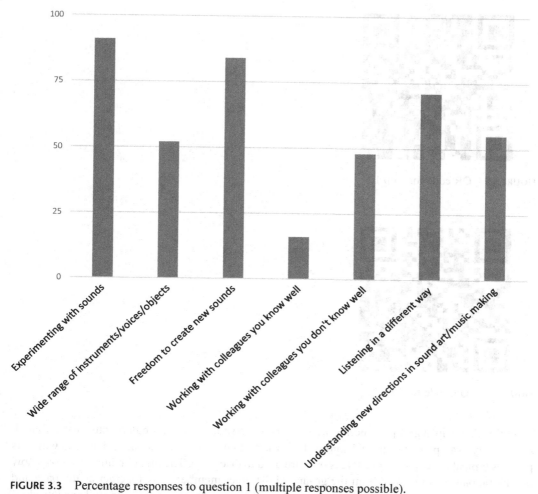

FIGURE 3.3 Percentage responses to question 1 (multiple responses possible).

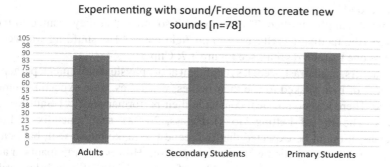

FIGURE 3.4 Percentage comparison between participants by age.

FIGURE 3.5 QR code for Clip 1.

FIGURE 3.6 QR code for Clip 2.

In Sweden, with a group including two dancers, the response was particularly physical, although the activity was preceded by a lengthy discussion about what was about to happen, which is perhaps typical for classical musicians who are used to discourse. The first clip initially shows how one participant breaks away from the group, whilst other members stay together as the pace and direction changes, including moving backwards. In the second performance, the students introduce crawling, rolling on the floor and jumping, saying the names of cards in a deck of playing cards.

The key finding here is that these students moved from a position of safety in the first performance to experimentation in the second.

The Brownies, by contrast, simply shuffled. They enjoyed it so much that they wanted to try a second time. As Loveless (the facilitator) reflects from his field notes: 'The shuffling was very good. I think it was the best one that we'd seen', see Figure 3.7: Clip 3.

The classroom music teachers, in contrast, were less confident about physical movement, perhaps because of the awkwardness of being asked to use their bodies, when they were expecting some kind of musical experience. There was no physical contact in their performance of '#1 Shuffle'.

The second work was 'The Joy of Letting Go' (Loveless, 2022a), see Appendix 2. For classically trained musicians, it might be expected that the request to make music with balloons would be rejected outright, or at least regarded with some scepticism. However, participants in all five groups and conference participants at the InMusic23 conference in July 2023 in Edinburgh enthusiastically took part, see Figure 3.8a: Clip 4a and Figure 3.8b: Clip 4b.

Is a balloon a musical instrument? More importantly, can a balloon make different sounds? What happens in an ensemble when participants explore what can be done with a balloon? It is

FIGURE 3.7 QR code for Clip 3 (Note: images have been pixilated for safeguarding reasons).

FIGURE 3.8a QR code for Clip 4a.

FIGURE 3.8b QR code for Clip 4b.

important to recognise that the facilitator left it open to players to create any sounds they wanted. So, for example, one might explore pitch, or rhythm, or call and response with people close by. This might be about blowing, or stretching the mouth of the balloon, or the speed of inflation. All the groups embraced the challenge enthusiastically.

The third work was 'Somewhere' (Loveless, 2022b), see Appendix 3. This is a vocal installation based on the opening phrase of 'Over the Rainbow' by Harburg and Arlen (1995) from the Hollywood musical *The Wizard of Oz*. The most interesting response to this activity came from the amateur choral society. As singers – with prior experience of choirs and some knowledge of notation – they knew the melody and were confident in their singing abilities

FIGURE 3.9   QR code for Clip 5.

FIGURE 3.10   QR code for Clip 6.

and chose to perform four times. Whilst they initially experimented with walking around and singing, they eventually decided to stand in a tight circle, see Figure 3.9: Clip 5. What emerged was that they started to listen in and understand that volume, intention and pitch could be combined in new ways without threatening their own musical identities. As one participant commented, 'This workshop gave me the opportunity to create different sounds through a group setting without any judgement. Exploring different art forms and how that can impact music and sound.'

Finally, workshop members took part in 'bloom' (Loveless, 2023), see Appendix 4. This work is a free-form exploration of sound involving a live processor created in Logic Pro with built-in plug-ins, a midi controller with 16 channels and a SM58 microphone and choice of about 400 musical instruments. The responses to this element of the workshop were very different.

The participants in Sweden incorporated dance into their performance, led musically by the pianist. The physical aspects of these two performances demonstrate that participants could make the work their own, whatever the circumstances and enjoyed the freedom to perform without judgement, see Figure 3.10: Clip 6. This, it may be argued, is important as it is the norm for conservatoire students to be assessed and judged regularly on their performances.

Surprisingly, the least successful session involved the classroom music teachers. One might assume that they have not only a background in music performance, but also an understanding of music pedagogy and the possible benefits of sound installation to promote inclusive music making in secondary schools with a diverse range of students. Of the 13 participants (six of whom provided feedback after the session), only four participated without any qualms. Using the school's

FIGURE 3.11   'bloom' with classroom teachers.

FIGURE 3.12   QR code for Clip 7.

assembly hall, participants could move outside or within the circle of instruments and playing positions. Whilst some participants clearly felt uncomfortable, one teacher grasped the purpose of movement in the workshop in their feedback: 'Understanding new directions in sound art/sound installation.'

Whilst there was some playful activity in the break with handbells which suggests that participants were experimenting freely, the classroom music teachers did not replicate these actions in the performances of 'bloom'. Whilst the facilitator encouraged performers to move around and interact with each other, it was mainly the Sixth Form students who explored different locations. Although all participants gave informed consent to be videoed and most completed feedback forms, there may be an aspect of feeling 'watched', which could illustrate a certain bewilderment with the process, as one teacher commented 'unusual but liberating' in describing the workshop.

In contrast, the Brownies, the youngest participants, approached the piece with enthusiasm and set their own parameters for performance, performing twice in a large group, see Figure 3.12: Clip 7. They subsequently decided to play in small ensembles for the last performance, demonstrating that they had fully understood that they were their own agents. As one commented in the feedback

FIGURE 3.13 QR code for Clip 8.

questionnaire, 'I love how we got together to make a beat. No matter what we use it sounds beautiful'.

The participants in Sweden again discussed what they planned to do in performing 'bloom'. Their performance appears to be led rhythmically by the postgraduate piano student with a considerable physical performance from the two dancers, as Figure 3.13: Clip 8 demonstrates. This, it could be argued, suggests the value of allowing conservatoire students to work freely, without any concerns about being judged for what they do. As one commented, 'This workshop gave me the opportunity to create different sounds through a group setting without any judgement'. Furthermore, the value of collaboration between physical performance students and musicians was also valued, 'As a person who wants more experience with cross collaboration with other disciplines in my UG/PG module, I think it would fit this setting as it connects artists from different fields together, wish I could have had this during my study'.

3.11 Discussion

The analysis of these five workshops reveals interesting differences, based on the questions below:

How do musicians participate in four different scenarios in a sound installation?
What implications does an approach based on sound installation have for pedagogy?
How does an inclusive, participatory workshop foster skills development in composition, improvisation and performance?

Sound art presupposes an affordance with objects in everyday life to make sound, (Strachan 2017). In order to challenge the hegemony of traditional approaches to composition and performance, there must be a fundamental understanding that music is for everyone, whatever their musical background or musical identity (Kanellopoulos, 2011; Niknafs, 2013). Interestingly, both the Swedish participants – classically trained and studying at a conservatoire – and the Brownies were the most dynamic in their performances. Why is this so? Whilst there is an assumption in some conservatoires that to play Cage's *4'33* you need to have an MA in music performance, these workshops demonstrate that a wide range of participants were prepared to explore sound and musical instruments in a free space. By doing so, some started to listen critically and 'in the moment' adapt their musical gestures to those around them, echoing Woods (2022, p.172), who states that 'creative practices emerge not within an individual, but in conversation with individuals, artefacts, ideas and contexts'.

The reluctance of some of the classroom music teachers to participate actively could be ascribed to a multitude of reasons. These might include unfamiliarity with the pedagogical and performative approach, fears about improvisation, a feeling that they were being 'watched' by the video camera or potentially being judged by the head of the borough music service who was a participant herself in the workshop. A further concern may have been anxiety about what this might mean for their professional practice. However, as Figure 3.11 illustrates, the musicians were loosely organised in a circle. Many who were seated may have felt uncomfortable about moving around playing their instruments. Nevertheless, there were moments of 'playfulness'.

The response of the Swedish participants reflects the experience of experimental musicians in the US documented by Woods (2019). They needed to discuss and conceptualise how they were going to respond to each work. These discussions gave them the confidence to step outside their comfort zones and experiment in performance, which was particularly evident in '#1 Shuffle' and 'bloom'.

The primary school students were overwhelmingly enthusiastic about the workshop. They demonstrated a joy in experimentation reflecting the installations conducted by Davenport et al., (2017). As one member of the group commented in the feedback questionnaire; 'I would like this in music lessons because theory is boring. I really enjoyed it!' So, if the majority of participants were happy to 'have a go' at something they had never experienced before, what are the implications for pedagogy?

If school schemes of work are traditionally based on notation and the process of developing e.g. manual dexterity in playing a keyboard, how does this address the needs of students with no musical background entering secondary education? Whilst there is an enthusiastic group of music educators in the US exploring pedagogical approaches to experimental music (Niknafs, 2013; Hickey et al., 2016, Sordahl, 2013 and Tinkle, 2015), such an approach raises immediate concerns about how such experimental works might be assessed. Moving back to the origins of experimental music, it is useful to recall Woods' definition (2019, p.459) encompassing experimental forms of notation, the use of objects, free improvisation and indeterminacy. Furthermore, it is important to recognise, as Woods states, quoting Gilmore (2014, p.183), that it is necessary to 'acknowledge and challenge the ongoing problematics of genre work that codify white supremacy and sexism within experimental music'.

Experimental music pedagogy should, it could be argued, be open to all, whatever their musical background, culture, language or ethnicity. This view is endorsed by Olson (2005, p.56), writing about community music making, describing the venue for rehearsal and performance as an emancipatory learning centre.

The video recordings clearly show that primary and secondary school students were able to explore sound and thus create their own works, although they may not have described these as 'compositions'. Indeed, the secondary students were wowed by the range of instruments and objects on offer – about 400 – and enjoyed playing or making sounds on instruments they had never had access to before. Some of the classroom music teachers had similar experiences: in a context without judgements, some tried new instruments or returned to instruments they had played in their youth, all of which is positive.

There are some small but interesting insights into the experiences of neurodivergent students. In the current study, one student had a diagnosis of attention deficit hyperactivity disorder [ADHD]. She participated twice in the workshop and demonstrated that she could play, improvise and perform without any fear of being judged by her peers or the participating classroom music teachers. Another student diagnosed with Autistic Spectrum Disorder [ASD] who specialises in music technology and recording found the workshop liberating and began to experiment with making sounds with instruments and objects, although she would not describe herself as a practicing musician. Wright (2021) examines the affordance of sonic-play instruments for non-verbal autistic students and it could be argued that a combination of acoustic instruments and music technology is a valuable approach to foster musical creativity for neurodivergent students.

3.12 Conclusion and Areas For Further Research and Action

How, therefore, can music educators adapt their pedagogy to encourage wider participation in musical composition and performance in secondary schools? The researchers would advocate an improvisatory, no holds barred approach to music composition, improvisation and performance. If music educators are to grow an inclusive, participatory, enjoyable and creative approach to musical learning, there is a need to offer an alternative to the old pedagogies of composition and reading notation. What would it be like if ten- and eleven-year-olds could explore sound as a means of personal expression? What would the implications be for examinations at KS4? How would examination boards react to a sound installation for a GCSE composition, which might include aspects of music technology? Such approaches challenge the very essence of music education in the UK. Is the music examiner – an organ scholar, or classically trained singer – equipped to assess a piece of sound art? How do assessment criteria need to be changed? What are the implications for neurodivergent students? Some of the participants in this pilot study had a diagnosis of ADHD and ASD and thoroughly enjoyed the workshops.

Our findings indicate that sound installations may have positive benefits for a wide range of student musicians in school settings and it would be logical to offer training to music educators to gain awareness of pedagogical approaches based around sound art and sound installation in order to widen participation in music. More research needs to be conducted at all stages of learning – HE, secondary schools, communities – in order to determine the most efficacious approach. There is clearly need for change. 'I want to break free' as Freddie Mercury (1992) sang. It is time to break free from the old ways and embrace a new pedagogy of composition, improvisation and performance which is equitable for all, whatever their musical background.

References

Barber-Kersovan, A. and Kirchberg, V. (2021) Free Ensembles and Small (Chamber) Orchestras as Innovative Drivers of Classical Music in Germany. In R. Hepworth- Sawyer, J. Paterson and R. Toulson (Eds), *Innovation in Music: Future Opportunities*. New York: Routledge Academics, pp. 290–305.

Bath, N., Daubney, A., Mackrill, D. and Spruce, G. (2020) The declining place of music education in schools in England. *Children and Society*, Vol. 34, pp. 443–457.

Beckstead, D. (2001) Will technology transform music education? *Music Educators Journal*, Vol. 87, No. 6, pp. 44–49.

Bell, C. L. (2018) Critical listening in the ensemble rehearsal. *Music Educators Journal*, March, Vol. 104, No. 3, pp. 17–25.

BERA (2018) Ethical Guidelines for Educational Research. Available at: www.bera.ac.uk/publication/ethical-guidelines-for-educational-research- 2018 [Accessed 23 10 2022].

Berkley, R. (2001) Why is teaching composing so challenging? A survey of classroom observation and teachers' opinions. *British Journal of Music Education*, Vol. 18, No. 2, pp. 119–138.

Bolden, B., Corcoran, S. and Butler, A. (2021) A scoping review of research that examines El Sistema and Sistema-inspired music education programmes. *Review of Education*, Vol. 9, e3267. https://doi.org/10.1002/rev3.3267

Butterworth, A. (1999). *Harmony in Practice*. London: ABRSM.

Cobussen, M., Meelberg, V. and Tmax, B. (2017) *The Routledge Companion to Sounding Art*. Abingdon: Routledge.

Cook, N. (2001) Between process and product: Music and/as performance. *Music Theory Online*. Available at www.mtosmt.org/issues/mto.01.7.2/mto.01.7.2.cook.html [Accessed 13 1 2023].

Cooper, B. (2019) *Primary Colours: The Decline of Arts Education in Primary Schools and How It Can Be Reversed*. Fabian Policy Report. London: The Fabian Society.

Corey, J. (2017) *Audio Production and Critical Listening: Technical Ear Training*. Abingdon: Routledge.

Cuervo, L. (2018) Study of an interdisciplinary didactic model in a secondary education music class. *Music Education Research*, Vol. 20, No. 4, pp. 463–479.

Davenport, J., Lockie, M. and Law, J. (2017) Supporting creative confidence in a musical composition workshop: Sound of colour. In: CHI PLAY, 15–18 October 2017, Amsterdam. Available at: https://clok.uclan.ac.uk/20430/ [Accessed 16 March 2023].

DfE, (2018) *Guidance on the English Baccalaureate*. Available at: www.gov.uk/government/publications/english-baccalaureate- ebacc/english-baccalaureate/ebacc [Accessed April 29 2020].

Durrant, C. and Welch, G. F. (1995) *Making Sense of Music: Foundations for Music Education*. London: Cassell.

Esslin-Peard, M. (2024) How Did We Learn to Create this Performance? A Tutor Models Student Reflective Practice? In C. Wiley and P. Gouzouasis (Eds), *The Routledge Companion to Music, Autoethnography and Reflexivity*. Abingdon: Routledge.

Folkestad, G. (2022) Researching Music and Music Education: The future in retrospect. *Keynote SEMPRE 50th Anniversary Conference*. Video available at: https://sempre.org.uk/50th [Accessed 15 October 2022].

Gibson, J. J. (1979) *The Ecological Approach to Visual Perception*. New York: Psychology Press.

Gilmore, B. (2014) Five Maps of the Experimental World. In D. Crispin and B. Gilmore (Eds) *Artistic Experimentation in Music: An Anthology*. Leuven: Leuven University Press, pp. 23–30. Available at: https://muse.jhu.edu/book/39256.

Gouzouasis, P. and Bakan, D. (2011) The future of music making and music education in a transformative digital world. Available from: www.unescoejournal.com/wp-content/uploads/2020/03/2-2-12- GOUZOUASIS.pdf [Accessed 31 March 2023].

Grant, J., Matthias, J. and Prior, D. (2021) *The Oxford Handbook of Sound Art*. New York, NY: Oxford University Press.

Hallam, S., Creech, A., Sandford, C., Rinta, T. and Shave, K. (2008) Survey of Musical Futures: a report from Institute of Education University of London for the Paul Hamlyn Foundation. Available at: https://discovery.ucl.ac.uk/id/eprint/1507448/1/Hallam2008_Musical_Futures_rep ort_final_version.pdf [Accessed March 10 2023].

Harburg, E. Y. and Arlen, H. (1995) *Over the Rainbow*. Arr. D. Nelson. Van Nuys, CA: Alfred Publishing, pp. 8–11.

Hickey, M., Ankney, K., Healy, D. and Gallo, D. (2016) The effects of group free improvisation instruction on improvisation achievement and improvisation confidence. *Music Education Research*, Vol. 18, No. 2, pp. 127–141.

Jeffreyes, B. (2018) Creative subjects being squeezed, schools tell BBC. Available at: www.bbc.co.uk/news/education-42862996 [Accessed 15 April 2022].

Kanellopoulos, P. A. (2009) Cage's short visit to the classroom: Experimental Music in music education – a sociological view on a radical move, in *Proceedings of the Sixth International Symposium on the Sociology of Music Education*, pp. 234–252. Available at: https://digital.library.unt.edu/ark:/67531/metadc1390657/m2/1/high_res_d/Kanell opoulosChapter.pdf [Accessed 18 December 2022].

Kanellopoulos, P. A. (2011) Freedom and responsibility: The aesthetics of free musical improvisation and its educational implications – A view from Bakhtin. *Philosophy of Music Education Review*, Vol. 19, No. 2, pp. 113–135.

Kitson, C. H. (1920) *Elementary Harmony Part 1*. Oxford: Oxford University Press.

Knowles, A. (1961) *Shuffle Event Score*. Available at: www.aknowles.com/eventscore.html [Accessed 2 January 2023].

Larsson, C. and Georgii-Hemming, E. (2019) Improvisation in general music education: A literature review. *British Journal of Music Education*, Vol. 36, No. 1, pp. 49–67.

London Borough of Hackney (2023) First Access programme. Available at: www.hackneyservicesforschools.co.uk/product/whole-class-tuition-new- instruments-year [Accessed 12 6 2023].

Loveless, S. D. (2022a) *The Joy of Letting Go*. Available at: www.samueldloveless.co.uk/iwanttobreakfree [Accessed 2 September 2022].

Loveless, S. D. (2022b) *Somewhere*. Available at: www.samueldloveless.co.uk/iwanttobreakfree [Accessed 5 October 2022].

Loveless, S. D. (2023) *bloom*. Available at: www.samueldloveless.co.uk/iwanttobreakfree [Accessed 10 October 2022].

Lovelock, W. (1939) *First Year Harmony*. London: A. Hammond.

Lovelock, W. (1949) *Free Counterpoint*. London: A. Hammond.

Macdonald, A. R. and Wilson, G. (2020) *The Art of Becoming: How Group Improvisation Works*. Oxford: Oxford University Press.

Mann, A. (1971) *The Study of Counterpoint from Johann Joseph Fux's Gradus ad Parnassum*. New York: W. W. Norton & Co.

Mercury, F. (1992) *I Want to Break Free*. *Queen Greatest Hits II*. London: Queen Music Ltd/EMI Publishing, pp. 113–135.

National Curriculum (2013) *National Curriculum in England: Music Programmes of Study*. Available at: www.gov.uk/government/publications/national- curriculum-in-england-music-programmes-of-study [Accessed 2 March 2023].

National Plan for Music Education (2022) Available at: https://assets.publishing.service.gov.uk/governm ent/uploads/system/uploads/attachment_data/file/1086619/The_Power_of_Music_to_Change_Lives.pdf [Accessed April 20 2022].

Niknafs, N. (2013) Free improvisation: What it is, and why we should apply it in our general music classrooms. *General Music Today*, Vol. 27, No. 1, pp. 29–34.

Nyman, M. (1974) *Experimental Music*. Cambridge: Cambridge University Press.

Olson, K. (2005) Music for community education and emancipatory learning. *New Directions for Adult and Continuing Education*, Vol. 107, pp. 55–64.

Reilly, N. (2021) A-level music education in schools could "disappear" in just over a decade. Available at: www.nme.com/news/music/a-level-music-education-in-schools-could-disappear-in-just-over-a-decade-3014383. [Accessed 2 February 2023].

Savage, J. (2018) Music Education for all. *FORUM*, Vol. 60 No. 1, pp. 111–122. doi: https//doi.org/10.15730/ forum.2018.60.1.111

Savage, J. and Challis, M. (2011) Electro-acoustic composition: Practical models of composition with new technologies. Available at: www.jsavage.org/jsorg/wp- content/uploads/2011/03/san.pdf [Accessed 26 November 2022].

Sawyer, K. R. (2008) Improvisation and teaching. *Critical Studies in Improvisation / Études critiques en improvisation*, Vol. 3, No. 2, pp. 1–4.

Sordahl, S. (2013) *Experiential Engagement in Experimental Music and Alternative Education*. M.F.A., Mills College. Available at: https://search-proquest- com.ezproxy.library.wisc.edu/docview/1366073603/ abstract/DF326C86E3C94DE APQ/1 [Accessed 9 March 2023].

Strachan, R. (2017) *Sonic Technologies, Popular Music, Digital Culture and the Creative Process*. London: Bloomsbury.

Steinitz, P. (1967) *Common Sense Harmony*. London: Mills Music.

Thomson, S. (2007) The pedagogical imperative of musical improvisation. *Critical Studies in Improvisation*, Vol. 3, No. 2, pp. 1–12.

Tinkle, A. (2015) Experimental music with young novices. *Leonardo Music Journal*, Vol. 25, No. 1, pp. 30–33.

Varese, E. and Wen-Chung, C. (1966) The liberation of sound. *Perspectives of New Music*, Vol. 5, No. 1, pp. 11–19.

Welch, G. F., Himonides, E., Saunders, J., Papageorgi, I., Rinta, T., Preti, C., Stewart, C., Lani, J. and Hill, J. (2011) Researching the first year of the National Singing Programme *Sing Up* in England: An initial impact evaluation. *Psychomusicology: Music, Mind and Brain*, Vol. 21, Nos. 1–2, pp. 83–97.

Whittaker, A. (2021) Teacher perceptions of A Level Music, tensions, dilemmas and decline. *British Journal of Music Education*, Vol. 38, pp 145–149.

Whittaker, A., Fautley, M., Kinsella, V. and Anderson, A. (2019) Geographical and social demographic trends of A Level music students. Available at: https://open- access.bcu.ac.uk/7511/1/RCM%20RAM%20 Report%20FINAL%20180419.pdf [Accessed 20 April 2023]

Winters, M. (2012) The challenges of teaching composition. *British Journal of Music Education*, Vol. 29, No. 1, pp. 19–24.

Woods, P. J. (2019) Conceptions of teaching in and through noise: A study of experimental musicians' beliefs. *Music Education Research*, Vol. 21, No. 4, pp. 459–468.

Woods, P. J. (2022) Learning to make noise: toward a process model of artistic practice with experimental music scenes. *Mind, Culture and Activity*, Vol. 29, No. 2, pp. 169–185.

Wright, J. (2021) Concepts for the Design of Accessible Music Technology. In R. Hepworth-Sawyer, J. Paterson and R. Toulson (Eds). *Innovation in Music*. [Online] Abingdon: Routledge.

Video Clips and Images

Photo 1: Still from video recording of 'bloom' with classroom music teachers. Recorded by the authors with informed consent from participants.

Video clip 1: Swedish participants in '#1 Shuffle'. Recorded by the authors with informed consent from participants.

Video clip 2: Swedish participants in '#1 Shuffle'. Recorded by the authors with informed consent from participants.

Video clip 3: Brownie group in '#1 Shuffle'. Recorded by the authors with informed consent from participants.

Video clip 4a: Conference presenters and participants at Innovation In Music 23 in Edinburgh performing 'The Joy of Letting Go'. Filmed and edited by August Liber Wlodzimierz, Vinephoto, (2023) and shared with kind permission of the videographer.

Video clip 4b: Brownies, Swedish participants and Ealing Choral Society in 'The Joy of Letting Go'. Recorded by the authors with informed consent from participants.

Video clip 5: Amateur choral society in 'Somewhere'. Recorded by the authors with informed consent from participants.

Video clip 6: Swedish participants in 'bloom'. Recorded by the authors with informed consent from the participants.

Video clip 7: Brownie group in 'bloom'. Recorded by the authors with informed consent from participants.

Video clip 8: Swedish participants in 'bloom'. Recorded by the authors with informed consent from participants.

Appendix 1

#1 Shuffle   Alison Knowles (1961)

The performer or performers shuffle into the performance area and away from it, behind, around, or throughout the audience. They perform as a group or solo: but quietly.

Premiered August 1963 at the National Association of Chemists and Performers in New York at the Advertiser's Club.

Appendix 2

The Joy of Letting Go   Samuel D Loveless (2022)

The Joy of Letting Go
a work for all

Score for conductor/facilitator;
to be explained or shown to the participants/performers.

Duration
Greater than two minutes.

Performance staging
Up to you! Have flare and be creative - embrace space, lighting and colourful balloons.

Performance method
For any number of people (suitable for one person or even large groups), each with a balloon of any size, colour and thickness. Biodegradable and/or recyclable balloons are preferred.

Performance notes
This piece focuses on the sound attributes of balloons as they are blown up, deflated, repeated, and eventually let go. As the process of blowing into the balloon and releasing the air repeats, the balloon will become more elastic, allowing its size to increasingly grow over the course of the piece.

Please keep all other unintended sounds to a minimum. No balloons should be popped during any given performance of this work - unless by accident.

After the performance, recycle any eco-friendly balloons and hang onto any balloons that aren't recyclable, as a memento or piece of art - or find a way to upcycle them, to provide even more joy after letting go.

Composer's note
This piece is designed to have a bit of fun and provide a little joy and laughter. Have fun and be creative!

SCORE

> A. Breathe in and blow into the ballon **once**. Then,
>
> B. Deflate the balloon in any fashion you like (quickly, slowly, in bursts), without piercing or popping it. *Don't let go of the balloon just yet!*
>
> ───────────── **Repeat** ─┘

C. When ready, release the balloon, allowing it to dance and twirl away. This could be when you're happy with the size of the balloon, it's almost ready to pop, you're tired or bored, or for some other reason.

Appendix 3

Somewhere   Samuel D Loveless (2022)

Samuel D. Loveless

Somewhere

for vocal ensemble/choir of four or more

Duration

Greater than five minutes.

Composer's note

A performance of 'Somewhere' should have a sense of pureness, innocence and hope at the heart of it.

Score

Harold Arlen's 'Over the rainbow' [main tune]

Execution

1. Individually sing ['ah' or 'hmm'] the main tune of Harold Arlen's 'Over the rainbow' (see transcription above).

   - Start when you wish or when directed to, in your preferred rhythm and tempo.

   - Sing and repeat this passage:

     i.   How you wish (i.e. rhythm, phrasing, dynamic, tone etc.)

     ii. As many times as you like

2. After a while or when directed, **you may** add harmonies, counter-melodies, pedals, ostinatos, questions/answers, and shouts as you wish.

   - Sing and repeat your newly added passage/phrase:

     i.   How you wish (i.e. rhythm, phrasing, dynamic, tone etc.)

     ii. Just once or as many times as you like

   - You may also sing, harmonise with and/or accompany the passages/phrases introduced by others.

3. When ready or when directed, repeat the passage/phrase you are singing for the final time, holding the last note (breathing as necessary).

   - Keep this note held until all ensemble members are doing the same, and continue to hold.

4. From this held chord, introduce a new passage/phrase or reintroduce a previously heard one.

   - Repeat this passage/phrase allowing the ensemble to become familiar with it, inviting them to sing, harmonise with, accompany, or create new passages/phrases from - just as the ensemble did with the initial 'Over the rainbow' melody.

   - Sing and repeat this passage:

     i.   How you wish (i.e. rhythm, phrasing, dynamic, tone etc.)

     ii. As many times as you like

   Repeat [steps 2-4];
   when ready continue to 'End'

End. When ready or when directed, repeat the initial 'Over the rainbow' melody once or more before coming to a rest.

   - Sing and repeat this passage:

     i.   How you wish (i.e. rhythm, phrasing, dynamic, tone etc.)

     ii. Just once or as many times as you like

   - Each performer should do this in their own time, coming off individually.

   - Once every ensemble member has done this, the piece has ended.

Somewhere
for vocal ensemble/choir of 8 or more

General performance instructions
The instructions below should be used to guide any given performance.

Within any performance:

i.   Start in the same key [as an ensemble], at any octave (with octave displacement).

ii.   Sing to 'ahh' or 'hmm', or a mixture of these throughout.

iii. Sing any given passage/phrase in your preferred rhythm and tempo.

iv.   Sing at any dynamic between *pppp* and *f,* keeping it the same or changing dynamics throughout.

v.   Sing any given passage/phrase as you wish (i.e. rhythm, phrasing, dynamic, tone etc.).

vi.   Repeat any given passage/phrase just once or as many times as you like.

vii. Breathe as and when you wish (i.e. at the ends of phrases or in the middle of notes etc.).

viii. Rest as and when you wish (i.e. at the ends of phrases or in the middle of notes etc.).

Once a passage/phrase has been introduced:

i.   It can be repeated endlessly by multiple members of the ensemble.

ii.   It may be heard continuously throughout, or stop and return at any point.

iii. You may sing, harmonise with and/or accompany it.

Be sure to listen throughout the performance, giving space to the other performers in the ensemble.

Manuscript paper (for optional use)
The staves below may be used for personal use to transcribe any additional musical material you may wish to sing.

Please note: although this work was originally intended for an a cappella ensemble/choir, it may also be performed by instruments, as well as instruments and voice.

Appendix 4

bloom · **Samuel D Loveless (2023)**

bloom
for ensemble of any instrumentation, size and musicial experience

Samuel D Loveless

Time

0'
20'–40'

At all times, listen; be conscious of the sounds created and aware of your unity as an ensemble

When ready, start to play

Create the smallest sounds you can
both individually and as an ensemble
allow these sounds to grow

—— *Repeat as you wish* ——

Contribute to a sonic bed, that:

1. acts as a foundation for musical ideas to grow out of
2. supports the sonic focal points

When ready, stop playing

—— *Repeat as you wish* ——

With the smallest change, make the largest impact

—— *Repeat as you wish* ——

Slowly, amalgamate your current sounds with those heard around you

—— *Repeat as you wish* ——

Create something new, anything you want
(there are no rules)

—— *Repeat as you wish* ——

4

REINVENTING THE MIXING DESK

A Comparative Review of the Channel Strip and the Stage Metaphor

Vangelis Katsinas

4.1 Introduction

Sound mixing heavily relies on the audio mixing interface (AMI), the mixing desk. The mixing desk was designed in the mid-20th century and evolved to host constituent modular channel strips that were influenced by the electronic elements available at that time. Despite technological developments in audio engineering, the mixing desk has remained largely unchanged to this day. Novel solutions could improve both usability and ergonomics, while also reducing the cognitive load imposed on the user.

This chapter presents a literature review, firstly of the Channel Strip (CS). It outlines the importance of usability in evaluating an interface and raises the question of whether sound engineers could benefit from a new AMI. Following that, it highlights the insufficient visual overview of the mix stemming from its inherent architectural design and examines the strictures imposed by its conventional controls.

The review then presents an alternative interface — the Stage Metaphor — in which individual sound sources are virtually visualised as musicians on stage. It focuses on the Stage Metaphor's intuitive design, its ability to provide an enhanced global understanding of a mix and explores its potential advantages over the CS paradigm. In addition, it addresses its problematic organisational overview and labelling of sound sources, as well as the lack of both intuitive physical controls and haptic feedback.

Finally, this chapter discusses the potential for AMI developers to take advantage of new technological advancements for the development of a novel AMI based on the Stage Metaphor, and how it might benefit from data visualisation principles, Dynamic Query filters, multi-modal interactive systems, and haptic technology.

4.2 Sound Mixing and 'Imaging'

In its most basic form, sound mixing is the activity of combining multiple sound sources, balancing their volume, and placing them in a stereo field in order to create an aesthetically pleasing whole. Mixing also involves equalisation, dynamic processing, and application of effects; however, in this chapter sound mixing will specifically refer to the aspects of balancing volume and stereo placement.

DOI: 10.4324/9781003396710-4

In the physical world, the distance and azimuth of a sound source dictate its apparent origin and help us localise it. The closer a sound source gets to the listener, the louder it becomes and vice-versa, known as the inverse square law. As the sound source moves across the horizontal axis, it reaches each ear at slightly different times (Interaural Time Differences) and at different levels (Interaural Level Differences).

When mixing, our mind creates a perceptual illusion of the apparent placement of sounds between the speakers. This process is called 'imaging' "because it is a figment of our imagination" (Gibson, 1997, p. 8). A mixing engineer might start mixing by visualising that image and then proceeding to reproduce it. This parallels the approach of a painter who first envisions an image before transferring it onto the canvas. Other times, they might instead create that image while mixing. Either way, the engineer refers to that mental image on an ongoing basis throughout the mixing process, as it enables them to gain a better understanding of the mix, including the spatial and dynamic relationship between the sound sources. This chapter will make further reference to this idea since beyond just the practical and aesthetic aspects of mixing, it also pertains to cognitive load and the way in which AMIs are able to relate to this image.

4.3   An Overview Of The Mixing Desk and The Channel Strip Paradigm

Mixing desks were introduced in the late 1930s by Western Electric and RCA (Liebman *et al.*, 2010; Scott, 2018). These first consoles were primarily used for motion picture production and were based on designs found in broadcasting, sound reinforcement, and telephone equipment. With the advent of stereo sound in the late 1950s the first mixing desks for music production were released (Ratcliffe, 2014), still heavily based on broadcasting consoles (Scott, 2018). Their design was rather minimal and primarily utilised three-inch rotary knobs, buttons, and switches. They did not feature phantom power, microphone preamplifiers, equalisation, effects, or dynamic processors. The relatively minimal signal processing performed via such desks was accomplished by routing signals to a limited range of outboard equipment, such as the Pultec EQ or Fairchild dynamic processor (Droney, 2003). A three-way switch control was utilised to pan sound to the left, centre or right (Melchior *et al.*, 2013), which was later replaced by the pan pot (panoramic potentiometer).

In the 1960s, major changes were made to the initial design. The shift from tube-based designs of the 1950s to solid-state electronics resulted in a substantial reduction in the size of mixing desks. Additionally, with the introduction of solid-state condenser microphones (previously, they had been vacuum-tube based), designers began to integrate microphone powering and mic preamplifiers. Bill Putnam was the first person to incorporate equalisers, compressors, cue sends, and echo returns and is credited for the layout of the mixing desk as we know it today (Swedien, 2003). Later in the '60s, Atlantic Records engineer Tom Dowd replaced the volume knob with the linear fader (Bell, Hein and Ratcliffe, 2015). This revolutionary development in audio engineering provided engineers with greater flexibility, enabling them to manipulate the volume of multiple channels with each hand. In Putnam's design, each audio signal was routed through a dedicated CS, "giving the user the ability to manipulate each audio channel in order to shape the overall mix" (Gelineck *et al.*, 2013, p. 2). Each CS comprised several knobs, switches, and a fader, each controlling a specific parameter. This one-to-one mapping of controls constitutes the CS paradigm. Up until that point, mixing desks were predominantly individually customised, and it was not until 1964 that Rupert Neve released the first commercial solid-state console (Scott, 2018).

Today, the CS can be found in analogue, digital, and software forms. The analogue mixing desk deploys a one-to-one mapping where every control has a dedicated function, making it both fast and intuitive. Digital mixing desks retain the traditional analogue mixing desk layout, but employ a one-to-many mapping, allowing a single control to be assigned to different parameters one at a time. They incorporate effects and built-in displays that provide visual feedback for various parameters and settings, offering greater flexibility, functionality, and detailed control. Furthermore, they allow for the design of compact-sized mixing desks by using layers and sub-menus. However, the one-to-many mapping and sub-menus do not offer a direct overview of the entire mixing desk, a feature most sound engineers find useful. As a result, the use of this design is more time-consuming, tedious, and "prone to unintended changes to audio controls" (Liebman *et al.*, 2010, p. 52). Additionally, it is less intuitive and tends to "present the user with a steeper learning curve" (Gelineck *et al.*, 2013, p. 2). Virtual mixing desks, found on nearly all Digital Audio Workstations (DAWs), web-based mixers, and iPads, also rely on the traditional real-world modelling of mixing desks (Mycroft and Paterson, 2011; Bell, Hein and Ratcliffe, 2015; De Man, Jillings and Stables, 2018; Gale and Wakefield, 2018) and carry the limitations of the analogue controls on to their digital counterparts (Théberge, 1997). The sound engineer can operate a virtual mixer using a mouse and keyboard, a touchscreen device, or a DAW controller.

Remarkably, despite the massive impact of technological developments in audio engineering, the mixing desk has remained largely unchanged since its birth (Gelineck, Büchert and Andersen, 2013; Gale and Wakefield, 2018) and continues to stand as the standard AMI to this day.

4.4   Challenges With The Channel Strip Metaphor

In the following section, a critical analysis of the CS will be undertaken to assess its adequacy and outline its limitations. Before delving into this enquiry, we need to question whether there really is a need to assess the mixing desk in the first place. After all, all the wonderful music we have ever listened to throughout the history of music production has been produced using the mixing desk. However, in order to evaluate an interface, one has to examine its usability, which encompasses efficiency, effectiveness, as well as user experience (UX). In addition, it is important to assess the cognitive load the interface imposes on the user, which refers to the mental resources required to operate the interface. Therefore, if there is a way to design an AMI that is faster, more precise, more intuitive, easier to learn and operate, more fun to work with, or an interface that allows the user to be more creative, then we should consider the replacement of the mixing desk.

4.4.1   Poor Overview of The Mix

Numerous researchers have questioned whether the CS paradigm serves as an optimal interface (Liebman *et al.*, 2010; Carrascal and Jordà, 2011; Cartwright, Pardo and Reiss, 2014; Ratcliffe, 2014; Wakefield, Dewey and Gale, 2017). Ratcliffe argues that although it offers precise control over various audio parameters in a mix, it faces a major challenge due to its architectural design (2014). This challenge, also highlighted by Diamante (2007), revolves around the mapping techniques utilised for panning and levelling, that prevent the sound engineer to directly visualise the mix as a whole.

In real life, panning is determined by the angle between the listener's head and the sound source. As the sound source moves along the horizontal axis — assuming the listener's head is in a fixed position — the panoramic position of the sound source changes. However, in the CS paradigm, the pan pot control remains static, rotating without any physical movement. Therefore, the pan pot has no direct correlation with the actual position of the sound source. Moreover, while the pan pot is in the centre position, the angle is accurate. However, as the pan pot moves towards either side, it becomes a rough approximation of the sound source's angle. When, for instance, an audio channel is panned hard left, the pointer on the pan pot is pointing to approximately 160°, while in reality the sound is coming from the front left speaker at approximately 45°, depending on the studio setup. Furthermore, the position of the channel on the mixing desk can be misleading with regard to its panoramic position in the mix (Ratcliffe, 2014). For example, a track that is panned hard right may actually be situated furthest left on the mixing desk.

The fader does not provide a direct mapping either. Its position is inversely proportional to the channel's volume level. The closer the fader is to the sound engineer, the farther away the sound source appears to be. Although users can adapt to the above mapping techniques, they impose additional cognitive load in processing this conceptual system. Another potential issue is that the fader position is not necessarily an absolute indicator of the perceived volume. This implies that, despite the fader levels suggesting one channel is louder than another, it could actually sound quieter. Such discrepancies stem from factors such as the level at which the tracks were recorded, compressed, and normalised, as well as their dynamic range and frequency content. This is not something the user can adapt to, and in such cases, the user has to accept that the volume mapping might be misleading when comparing fader levels between tracks.

To create a mental image of the mix based on the overview of the mixing desk, the user needs to scrutinise and memorise pan pot and fader positions across all channels. While it is easy to see where the sound is positioned relative to the speakers and the listening point on a single channel, this task becomes exceedingly challenging — if not impossible — as the number of channels increases. In fact, according to Miller (1956), the amount of information our working memory can simultaneously retain and process is limited to approximately seven items, give or take two. This underscores the limited number of fader and pan pot positions human memory can hold at any given time. Furthermore, processing a large amount of visual information imposes a significant cognitive load on sound engineers, diverting their attention from auditory tasks (Mycroft, Reiss and Stockman, 2013; Whitenton, 2013; Dewey and Wakefield, 2016). In that respect, the CS does not provide an efficient overview of the mix, which could reduce some of the cognitive load from the auditory modality.

4.4.2 Poor Ergonomics

Apart from the poor mix overview of the CS, Carrascal and Jordà (2011) argue that the outdated controls found on the mixing desk are "not necessarily ergonomical or adequate" (p. 100), which poses an additional challenge. With the advent of the fader, the 3-inch (approximately 76 mm) knobs became significantly smaller to match the width of the fader and minimise the overall size of the console. Research findings indicate that the optimal size for a knob in terms of ergonomics is approximately 50 mm (Øvergård et al., 2007). The standard knob found on the mixing desk is 6 mm. As a result, operating these small knobs sometimes becomes challenging, often forcing sound engineers to adopt awkward hand postures while making adjustments. Many researchers have found that it is possible for critical factors such as highly repetitive hand motions, high pinch force exertions and awkward hand postures to cause injuries and musculoskeletal disorders (Villanueva, Dong, and Rempel 2007; Rock, Mikat, and Foster 2001; Burke, Main, and Freeman 1997; Rolian, Lieberman, and Zermeno 2011; Ng et al. 2013a; Ng et al. 2013b, as cited in Tan et al., 2015). Faders can lead to wrist strain, which is why many engineers prefer faders that are parallel to the ground (Liebman et al., 2010). For shorter engineers, accessing certain parts and controls on a large mixing desk — particularly those located at the top — might be challenging. Buttons also pose issues. Sound engineers often struggle to determine whether a button is pressed or not unless they physically touch it, in order to haptically identify its state. This is very common with buttons associated with phantom power, low-cut filter, and other on/off switches, especially those positioned farther away from the engineer. Moreover, some of the buttons can be very small or hard to access, particularly when situated between other controls.

The small size and the close proximity of the controls also make it hard to discern the associated alphanumerical information. Alphabetical information is often abbreviated (e.g., LCF for Low Cut Filter, Pk for Peak, etc.), potentially causing difficulties for new users to understand. Numerical values are often approximate due to the limitations imposed by the compact layout, such as frequency values in the equalisation section. Furthermore, all controls maintain a fixed size and shape and cannot be customised to fit the user's preference. In terms of customisation, the traditional controls offer limited options beyond replacing a control with a different-coloured one.

4.4.3 Limited Haptic Feedback

The size and shape of the controls are important not only for ergonomical purposes but also for providing visual and haptic feedback. Shape and size are variables that can be used to graphically and haptically represent parameter information. In addition to their fixed size and shape, faders and knobs maintain a consistent resistance throughout their range of motion, which limits their ability to haptically convey information about their current value. The limited haptic feedback they offer merely allows users to determine whether a control has reached its maximum or minimum position when the control cannot physically move any further. Minimal haptic cues, such as a tactile dent at the centre of the pan pot, are used to indicate that the sound is equally distributed between the two monitors. Similarly, some knobs incorporate a dent to indicate unity gain. Buttons, on the other hand, provide a subtle tactile vibration or "click" as a confirmation that the user's action has been registered. However, this haptic feedback does not indicate whether the button has been pressed up or down.

The poor haptic design of the controls does not allow users to customise haptic feedback, such as adding extra dents, altering resistance, or adding haptic effects. In essence, the purpose of the controls is to merely adjust audio parameters, which makes the interaction with these controls less intuitive.

4.5 Stage Metaphor

4.5.1 Overview

To address the poor visual feedback provided by the CS paradigm, an alternative metaphor has been developed based on psychoacoustic principles that align with the way humans perceive and localise sound in the real world (Gelineck, Andersen and Büchert, 2013). The proposed paradigm is known as the Stage Metaphor (SM). The SM was originally introduced in the 1990s by Gibson (1997) as the 'virtual mixer', offering a more efficient way to visualise the mix. Although Gibson's interface was not intended to manipulate audio parameters, it inspired a series of related paradigms which allowed both the visualisation and control of the mix (Liebman et al., 2010; Carrascal and Jordà, 2011; Gelineck, Büchert and Andersen, 2013; Lech and Kostek, 2013; Ratcliffe, 2014; Dewey and Wakefield, 2016; Jordan et al., 2016; Dewey and Gale, 2017).

The fundamental idea of the SM is that each audio channel is graphically represented by a movable icon, known as a widget. Each widget is positioned on a stage, which is the space where the mix occurs and is related to the area between the speakers and the listener. The widget's position relative to a fixed virtual listening point determines the channel's volume and panning in the overall mix.

In contrast to the well-established CS which has a standardised layout, the SM can be found in various designs, utilising a 2D or 3D stage and a variety of different input controls. Although several GUIs (Graphical User Interfaces) and various visualisation styles have been designed and evaluated, there is currently no consensus on the most effective model.

4.5.2 Advantages of the Stage Metaphor Over the Channel Strip

There is compelling evidence that the SM offers improved usability compared to the CS, which encompasses efficiency, effectiveness, and UX. Research has demonstrated that interfaces based on the SM can enhance efficiency by significantly reducing the normalised task completion time (NTCT) in certain tasks (Carrascal and Jordà, 2011; Gelineck and Korsgaard, 2014). Moreover, results indicate that the SM provides a better overview of the mix and significantly improves concurrent critical tasks (Mycroft, Stockman and Reiss, 2015). Other research shows that the SM is generally perceived as a more desirable interface compared to the CS, leading to an overall enhancement of UX (Gelineck et al., 2013). Furthermore, the SM provides a more creative and exploratory environment for mixing (Lech and Kostek, 2013; Gelineck and Korsgaard, 2014), which suggests that the interface itself can influence the outcome of the mix. In addition, the SM is characterised by an intuitive design (Gelineck et al., 2015), which makes it easier for users to familiarise themselves with its functionality and reduces the inherent learning curve (Holladay, 2005).

But what is it that makes the SM more intuitive? Compared to the CS, "the conceptual system of the stage metaphor is much closer to the mental model of the user in how they conceive an overall mix" (Gelineck, Büchert and Andersen, 2013, p. 737). This implies that the mix overview in the SM aligns with the perceptual image created through the 'imaging' technique. Such alignment makes it far easier for the user to have a "clear visual overview of how audio channels contribute to the overall mix" (Gelineck and Korsgaard, 2014, p. 2), which has been found to "significantly improve concurrent critical listening tasks" (Mycroft, Stockman and Reiss, 2015, p. 682). Moreover, the authors argue that

> the enhanced global understanding of the mix that this provides may help users to quickly see patterns within the mix, avoid common errors such as masking or bunching of elements within a certain stereo position (Case, 2007) and allow any outliers (in terms of volume, pan etc.) to be easily attended and selected (Wolfe, Klempen and Dahlen, 2000).
>
> *(Mycroft, Stockman and Reiss, 2015, pp. 682–683)*

The user sets the positioning of the channels by moving the widgets, in a similar way to moving performers or sound sources in a physical space. This method is more natural and intuitive compared to using separate controls to set the volume and panning. While controlling the fader and pan pot feels like a remote control for adjusting the level and panning, interacting with the widgets feels like physically moving the actual sound rather than manipulating its attributes (Gelineck, Andersen and Büchert, 2013). This multiparametric approach also enables the user to simultaneously set the volume and panning of each channel. In contrast, in the CS both the fader and pan pot need to be adjusted separately to control the sound source's position. Lech and Kostek found that sound engineers appreciate the possibility of controlling two parameters of the same channel at the same time (2013).

4.6 Disadvantages of the Stage Metaphor

4.6.1 Clutter, Occlusion and Track Selection Time

The SM shows great potential as an alternative AMI, yet it comes with its own drawbacks. Due to the random distribution of the channels, widgets become cluttered as the number of channels increases (Dewey and Wakefield, 2016; Gelineck and Uhrenholt, 2016). Clutter occurs when too many widgets appear on the stage, making it difficult to interact with the widgets efficiently.

The overwhelming and oftentimes chaotic display of the stage when dealing with a large number of channels, reduces visual clarity and may also result in occlusion. Occlusion pertains to the partial or complete blocking of a widget when one or more of them overlap. This arises between tracks with similar panoramic positions and gain levels. While this visual masking might be useful in indicating possible audio masking (Diamante, 2007; Mycroft, Stockman and Reiss, 2015), it also makes it challenging for users to view and access hidden or partially obscured widgets. Clutter and occlusion are somehow related. Clutter occurs as the number of channels increases, whereas occlusion can occur as a result of clutter or even when as few as two audio channels occupy the same physical space on the stage.

Another issue with the channel organisation in the SM is the increased track selection time (TST), which refers to the time it takes to search the stage to identify and select the desired audio channel. Dewey and Wakefield argue that "intuitive track selection is a key AMI user requirement" (2017, p. 2). It has been found that increased TST may cause confusion and have a negative impact on the UX and the efficiency of the interface (Dewey and Wakefield, 2017). The problematic channel organisation and labelling of the SM are likely the reason why the CS has remained the predominant mixing metaphor to this day.

4.6.2 Lack of Efficient Input Controls

Most SM designs employ a 2D GUI controlled by a keyboard and mouse. 'Reactable' stands out as a distinctive implementation of the SM that utilises tangible objects for manipulating widgets. 'Reactable' is an electronic musical instrument based on a round translucent tabletop with a backlit surface, using smart tangibles to control sound parameters. These smart tangibles are tangible blocks with embedded sensors that can be moved, and their positions dynamically manipulate the sound in real time. Although it was initially conceived as a modular synthesiser, it inspired the design of a tangible AMI based on the SM (Gelineck et al., 2013). Other implementations include touchscreen interfaces (Diamante, 2007; Carrascal and Jordà, 2011), hand-gesture-based interfaces (Carrascal and Jordà, 2011; Lech and Kostek, 2013; Ratcliffe, 2014; Gelineck and Korsgaard, 2015), haptically augmented interfaces (Melchior et al., 2013; Gelineck and Overholt, 2015; Quiroz and Martin, 2020), and interfaces utilising virtual reality (VR) and game controls (Jordan et al., 2016; Gale and Wakefield, 2018).

A fundamental consideration during interface development is the way in which the user physically interacts with the interface (Gelineck, Büchert and Andersen, 2013). Interfaces controlled by a mouse and keyboard are disadvantaged compared to the mixing desk, since they exhibit insufficient ergonomics and are limited to controlling only one parameter at a time (Lech and Kostek, 2013). Moreover, they are overly dependent on the computer screen and lack efficient haptic feedback. The constant visual attention to the display has been found to increase cognitive load and to compromise the user's critical listening ability (Lech and Kostek, 2013).

Tangible user interfaces (TUIs) present a potential alternative AMI since they provide haptic feedback and minimise dependence on the screen, enabling users to focus more on mixing. Moreover, the direct control of digital representations with tangibles encourages interaction with both hands for parallel input (Kirk *et al.*, 2009). However, TUIs come with some significant limitations. With TUIs it is not possible to place two tangibles in the exact same position, and too many tangibles result in clutter. Most importantly, tangibles are static and cannot change appearance (Kirk *et al.*, 2009). Therefore, useful variables such as motion, speed, shape, size, texture and so on, cannot be used to convey information about the mix. Moreover, TUIs may occupy a lot of physical space, which is not ideal for the context of professional mixing (Carrascal and Jordà, 2011).

Touchscreen interfaces also offer the advantage of direct manipulation of visual objects. Preliminary data suggests that they can potentially reduce the learning curve, enhance creativity and that they are more preferable and time-efficient compared to the analogue mixing desk (Carrascal and Jordà, 2011). However, they suffer from poor ergonomics and lack haptic feedback (ibid.), whereas other studies indicate that they are susceptible to 'exit error' (Gelineck, Büchert and Andersen, 2013; Gale and Wakefield, 2018). 'Exit error' refers to the unintended change of a parameter caused by micro-movements of the hand as the finger is lifted from the widget. Consequently, this characteristic makes touchscreen interfaces less suitable for tasks requiring fine-tuning (Gelineck *et al.*, 2013).

Hand-gesture interfaces allow for eye-free interaction with the interface and show great potential, but lack precision and are prone to 'exit error' and 'gorilla arm' fatigue (Gelineck and Korsgaard, 2015). The term 'gorilla arm' refers to the fatigue and discomfort that users experience in their hands when performing mid-air hand-gestures for extended periods. Although certain gesture-based input controls appear preferable compared to the conventional controls, as well as the mouse and keyboard, research has shown that they can increase cognitive load (Quiroz and Martin, 2020). Other gesture-based interfaces have been reported to be inconsistent, hard to use, frustrating, and time-consuming (Wakefield, Dewey and Gale, 2017). While some studies suggest that these types of interfaces are more ergonomic compared to those using a mouse and keyboard (Lech and Kostek, 2013), other findings point to ergonomic challenges (Gelineck and Korsgaard, 2015). These conflicting results may stem from variations in the input devices employed within the interfaces.

Promising preliminary data from haptically augmented interfaces employing the Novint Falcon device show that a haptic device may offer faster and more preferable control compared to the mouse and keyboard (Melchior et al., 2013; Quiroz and Martin, 2020) and reduce dependency on looking at the screen (Gelineck and Overholt, 2015). The Novint Falcon is an affordable 3D haptic controller that provides force feedback, enabling users to feel the shape, weight, and texture of virtual objects. The incorporation of these haptic variables allows the haptic representation of volume data, a concept known as 'volume haptisation' (Iwata and Noma, 1993). However, although the Novint Falcon excels in facilitating 3D movements, a mouse-based interface allows for more precise movements in a 2D environment (Melchior *et al.*, 2013). As haptic interfaces continue to advance, they may eventually overcome their current limitations and emerge as standard input controls in interfaces that stand to benefit from haptic augmentation.

Finally, Lech and Kostek argue that game controllers, infrared sensors, or accelerometers do not provide adequate ergonomics for adoption in AMIs (2013).

4.7 Future Work

New technological developments present numerous possibilities for advancing AMIs. Although both the CS and SM can benefit from such technologies, the CS is restricted by its inherent architectural inability to provide a sufficient overview of the mix. The SM presents a lot more potential due to its intuitive design, efficient mix overview, and ability to reduce cognitive load. However, further work is required to determine both the quantity and organisation of information presented to the user, through data visualisation and Dynamic Query (DQ) filters. Data visualisation involves the use of advancing graphic technology to minimise the time and effort of finding and accessing visual information (Robertson, Card and Mackinlay, 1993). This technology employs dynamic information, taking advantage of the user's cognitive ability to detect changes in colour, shape, size and texture (Shneiderman *et al.*, 2010). According to Mendelzon (1996), many researchers have proposed data visualisation as "a technology to address the severe problems of disorientation and information overload" that arise when interacting with a vast, random and disorderly informational space (p. 13). Furthermore, future studies should explore the implementation of more sophisticated DQ filters to temporarily simplify the GUI of the SM. DQ filters is an interactive mechanism that allows the user to dynamically filter displayed information based on specific criteria in real-time. Examples of criteria could include dynamic range, volume, group category, panoramic range, frequency spectrum, etc. Prior research has demonstrated that DQ filters can significantly improve visual search and critical listening tasks (Mycroft, Stockman and Reiss, 2016). Data visualisation and DQ filters offer potential solutions to address the problematic organisational overview of the SM, namely clutter, occlusion, and TST. Dewey and Wakefield (2014) suggest that AMI development should be guided by existing research in Human Computer Interface (HCI) design . The aim of HCI design is to inform the development of interfaces that enable users to achieve their goals effectively (Norman, 2002).

HCI designers optimise UX by minimising cognitive load and making the interaction with the interface more intuitive and efficient. Audio mixing can be a high-workload, multi-task activity that requires significant cognitive resources. The multitasking nature of audio mixing poses a challenge for potential cognitive overload. Although there is no way to entirely eliminate cognitive load, it is, however, important that there are enough mental resources available for making critical listening decisions. A successful AMI is one that allows the user to be creative and is designed in such a way as to minimise cognitive overload. Wickens' multiple resource theory suggests that distributing the load across multiple modalities — the auditory, visual, and haptic — will improve performance by reducing cognitive load (Wickens, 2002; Mayer and Moreno, 2003; Prewett *et al.*, 2006). Given that the auditory modality is primarily occupied with critical listening tasks and the visual modality processes information from the DAW, the haptic channel becomes a viable option to interact with the AMI. Integrating the haptic channel will add a third sensory dimension and lead to the development of a tri-modal AMI, where the haptic channel is used not only to input information through the controls but also to provide information about the mix through haptic feedback.

Haptics is the field of study that explores the sense of touch and its role in human perception and interaction with the world. It also refers to the technology associated with the sense of touch in virtual environments and enables the user to haptically interact with virtual objects. Haptics has significant potential to enhance the usability of an AMI and reduce cognitive load. According to Avila and Sobierajski (1996), the addition of haptic cues can be "particularly useful when the user attempts to precisely locate a feature within a volume, or to understand the spatial arrangement of complex three-dimensional structures" (p. 197). Other work has shown that haptic feedback can improve NTCT (Hasser *et al.*, 1998) and the sense of presence and awareness when added to a visual interface (Lee and Kim, 2008). A meta-analysis of 43 studies on the effects of multi-modal feedback on user performance showed that adding tactile feedback to visual feedback was found to improve performance overall, reduce reaction times, and provide a significant advantage over using a visual-only feedback system (Burke *et al.*, 2006), especially when participants were performing multiple tasks with high workload (Prewett *et al.*, 2006). Moreover, the addition of tactile feedback improves target acquisition, which involves searching the environment to identify a target (Prewett *et al.*, 2006). This could potentially improve TST. Although these are preliminary data from the evaluation of haptic devices, future research should focus on exploring and evaluating more recent haptic devices. Additionally, more work is needed to optimise haptic effects for audio mixing, investigate ways haptics can effectively convey information about audio parameters, and inform haptic device developers of haptic features that could benefit AMIs.

Finally, numerous researchers have suggested that alternative input controls to faders, knobs, and buttons should be considered (Lech and Kostek, 2013; Ratcliffe, 2014; Wakefield, Dewey and Gale, 2017; Quiroz and Martin, 2020; Gelineck *et al.*, 2013). Modern technology, such as novel hardware controls, haptics, extended realities, etc., opens up new ways for interacting with interfaces and offers controls that can be ergonomically and technically superior. These controls could potentially convey haptic or graphic information about volume, pan, frequency, and other audio parameters.

4.8 Conclusion

This paper has provided an extensive literature review of the CS and has critically analysed the limitations stemming from its architectural design and outdated electronic components. Several researchers have questioned whether the mixing desk is an adequate AMI (Liebman *et al.*, 2010; Carrascal and Jordà, 2011; Cartwright, Pardo and Reiss, 2014; Ratcliffe, 2014; Wakefield, Dewey and Gale, 2017), while others have emphasised that interfaces should be evaluated based on usability (Dewey and Wakefield, 2014) and the cognitive load they impose on the user (Whitenton, 2013). Although the CS offers precise control over many audio parameters, it does not provide the user with a direct overview of the mix. Moreover, the traditional controls lack ergonomics (Carrascal and Jordà, 2011) and offer limited haptic feedback.

As an alternative to the CS, the SM has been proposed, taking inspiration from the way humans perceive sound in the real world. A comprehensive review and a critical assessment of the advantages of this paradigm over the CS has been given. In particular, its ability to simulate 'imaging' and enable users to visually develop an enhanced spatial understanding of the mix has been underlined. Although the SM offers enhanced usability and a more intuitive design compared to the CS, it comes with its own challenges. Clutter, occlusion and slow TST are some of the main constraints imposed by its problematic organisational overview. These limitations are likely the reason that the CS still remains the primary AMI (Gelineck, Büchert and Andersen, 2013). Further work is required to address these drawbacks. Research from HCI can inform the development of novel AMIs to maximise usability and minimise cognitive load. More sophisticated data visualisations and DQ filters need to be explored to address the problematic organisational design of the SM. In addition, according to Wickens' multiple resource theory, integrating the haptic modality may reduce the cognitive load of an interface (Wickens, 2002; Prewett *et al.*, 2006). Haptic integration can potentially improve usability and might be the key feature in further improving the SM (Burke *et al.*, 2006; Prewett *et al.*, 2006; Gelineck and Overholt, 2015).

References

Avila, R. S. and Sobierajski, L. M. (1996) 'A haptic interaction method for volume visualization', in *Proceedings of Seventh Annual IEEE Visualization'96*. San Francisco, CA: IEEE, pp. 197–204. doi: 10.1109/VISUAL.1996.568108

Bell, A., Hein, E. and Ratcliffe, J. (2015) 'Beyond skeuomorphism: The evolution of music production software user interface metaphors', *Journal on the Art of Record Production*, 9. Available at: www.arp journal.com/asarpwp/beyond-skeuomorphism-the-evolution-of-music-production-software-user-interf ace-metaphors-2/ (Accessed: 14 July 2022).

Burke, J. L. *et al.* (2006) 'Comparing the effects of visual-auditory and visual-tactile feedback on user performance: A meta-analysis', in *Proceedings of the 8th International Conference on Multimodal Interfaces*. Banff, Alberta, Canada, pp. 108–117. doi: 10.1145/1180995.1181017

Carrascal, J. P. and Jordà, S. (2011) 'Multitouch interface for audio mixing', in *Proceedings of the International Conference on New Interfaces for Musical Expression. New Interfaces for Musical Expression, NIME*, Oslo, Norway, pp. 100–103. doi: 10.5281/zenodo.1177983

Cartwright, M., Pardo, B. and Reiss, J. D. (2014) 'Mixploration: Rethinking the audio mixer interface', in Proceedings of the 19th international conference on Intelligent User Interfaces. Haifa, Israel, pp. 365–370. doi: https://doi.org/10.1145/2557500.2557530

Case, A. (2007) *Sound FX: Unlocking the Creative Potential of Recording Studio Effects*. New York: Routledge. doi: 10.4324/9780080548968

De Man, B., Jillings, N. and Stables, R. (2018) 'Comparing stage metaphor interfaces as a controller for stereo position and level', in *4th Workshop on Intelligent Music Production*. Huddersfield, United Kingdom. Pages 3, Available at: www.brechtdeman.com/publications/pdf/WIMP4.pdf (Accessed: 28 June 2022) https:// scholar.google.com/scholar?hl=en&as_sdt=0%2C5&q=Comparing+stage+metaphor+interfaces+as+a+ controller+for+stereo+position+and+level&btnG=.

Dewey, C. and Wakefield, J. (2014) 'A guide to the design and evaluation of new user interfaces for the audio industry', in *Audio Engineering Society Convention 136*, Berlin, Germany: Audio Engineering Society, p. 10. Available at: www.aes.org.ezproxy.uwl.ac.uk/e-lib/inst/browse.cfm?elib=17218 (Accessed: 3 June 2022).

Dewey, C. and Wakefield, J. (2016) 'Novel designs for the audio mixing interface based on data visualization first principles', in *Audio Engineering Society Convention 140*. Paris, France: Audio Engineering Society. Available at: www.aes.org/e-lib/browse.cfm?elib=18264 (Accessed: 8 June 2022).

Dewey, C. and Wakefield, J. (2017) 'Formal usability evaluation of audio track widget graphical representation for two-dimensional stage audio mixing interface', in *Audio Engineering Society. 142nd Audio Engineering Society International Convention 2017*, Berlin, Germany: Audio Engineering Society, p. 11. Available at: www.aes.org/e-lib/browse.cfm?elib=18672 (Accessed: 23 June 2022).

Diamante, V. (2007) 'Awol: Control surfaces and visualization for surround creation', *University of Southern California, Interactive Media Division, Tech. Rep.* Available at: www.surroundshape.com/diamante_awol _thesispaper_18.pdf (Accessed: 21 July 2022).

Droney, M. (2003) *Mix Masters: Platinum Engineers Reveal Their Secrets for Success*. United States of America: Berklee Press.

Gale, W. and Wakefield, J. (2018) 'Investigating the use of virtual reality to solve the underlying problems with the 3D stage paradigm', in *Proceedings of the 4th Workshop on Intelligent Music Production*. United Kingdom, Huddersfield, pp. 1–4. Available at: https://research.hud.ac.uk/media/assets/document/research/ 2-WillGale-WIMP2018-VR.pdf (Accessed: 21 September 2022).

Gelineck, S., Andersen, J. and Büchert, M. (2013) 'Music mixing surface', in *Proceedings of the 2013 ACM international conference on Interactive tabletops and surfaces*. New York, NY: Association for Computing Machinery, pp. 433–436. doi: 10.1145/2512349.2517248

Gelineck, S., Büchert, M. and Andersen, J. (2013) 'Towards a more flexible and creative music mixing interface', in *CHI'13 Extended Abstracts on Human Factors in Computing Systems*. Paris, France: Association for Computing Machinery, New York, United States, pp. 733–738. Available at: https://dl.acm.org/doi/abs/ 10.1145/2468356.2468487 (Accessed: 12 July 2022).

Gelineck, S. and Korsgaard, D. (2014) 'Stage metaphor mixing on a multi-touch tablet device', in *Audio Engineering Society Convention 137*. Los Angeles, CA: Audio Engineering Society. 10 pages, Available at: www.aes.org/e-lib/browse.cfm?elib=17456 (Accessed: 17 September 2022).

Gelineck, S. and Korsgaard, D. (2015) 'An exploratory evaluation of user interfaces for 3D audio mixing', in *138th AES Convention. Audio Engineering Society Convention 138*, 4 pages, Warsaw, Poland: Audio Engineering Society. Available at: www.aes.org/e-lib/browse.cfm?elib=17636 (Accessed: 7 December 2022).

Gelineck, S., Korsgaard, D. M. and Büchert, M. (2015) 'Stage-vs. channel-strip metaphor: Comparing performance when adjusting volume and panning of a single channel in a stereo mix', in *Proceedings of the International Conference on New Interfaces for Musical Expression (NIME 2015)*. Baton Rouge, LA: Louisiana State University, pp. 343–346. Available at: https://vbn.aau.dk/en/publications/stage-vs-channel-strip-metaphor-comparing-performance-when-adjust (Accessed: 19 May 2023).

Gelineck, S. and Overholt, D. (2015) 'Haptic and visual feedback in 3D audio mixing interfaces', in *Proceedings of the Audio Mostly 2015 on Interaction With Sound*. Article No. 14, pp. 1–6. Thessaloniki, Greece: Association for Computing Machinery (AM '15). doi: 10.1145/2814895.2814918

Gelineck, S. and Uhrenholt, A. K. (2016) 'Exploring visualisation of channel activity, levels and EQ for user interfaces implementing the stage metaphor for music mixing', in *2nd AES Workshop on Intelligent Music Production*. 3 pages. London, UK: Audio Engineering Society. Available at: https://vbn.aau.dk/ws/files/241846470/Gelineck_AES_WIMP_16.pdf (Accessed: 28 November 2022).

Gelineck, S. *et al.* (2013) 'Towards an Interface for Music Mixing based on Smart Tangibles and Multitouch', in *NIME 2013*. Daejeon and Seoul, Korea Republic: Association for Computing Machinery, p. 7. Available at: www.researchgate.net/publication/274660638_Towards_an_Interface_for_Music_Mixing_based_on_Smart_Tangibles_and_Multitouch (Accessed: 14 June 2022).

Gibson, D. (1997) *The Art of Mixing: A Visual Guide to Recording*. 1st Edition. Vallejo, California: MixBooks (Mix pro audio series).

Hasser, C. J. *et al.* (1998) 'User Performance in a GUI Pointing Task With a Low-Cost Force-Feedback Computer Mouse', in *ASME International Mechanical Engineering Congress and Exposition. Dynamic Systems and Control*, Anaheim, California: American Society of Mechanical Engineers, pp. 151–156. doi: 10.1115/IMECE1998-0247

Holladay, A. (2005) 'Audio dementia: A next generation audio mixing software application', in *Audio Engineering Society Convention 118*. Barcelona, Spain. 6 pages. Available at: www.aes.org/e-lib/browse.cfm?elib=13103 (Accessed: 12 July 2022).

Iwata, H. and Noma, H. (1993) 'Volume haptization', in *Proceedings of 1993 IEEE Research Properties in Virtual Reality Symposium*. San Jose, CA: IEEE, pp. 16–23. doi: 10.1109/VRAIS.1993.378268

Jordan, D. *et al.* (2016) 'Spatial audio engineering in a virtual reality environment', *Gesellschaft für Informatik e.V.* doi: http://dx.doi.org/10.18420/muc2016-mci-0217

Kirk, D. *et al.* (2009) 'Putting the physical into the digital: Issues in designing hybrid interactive surfaces', in *Proceedings of British HCI 2009. People and Computers XXIII Celebrating People and Technology*, Churchill College Cambridge: Cambridge University Press. doi: 10.14236/ewic/HCI2009.5

Lech, M. and Kostek, B. (2013) 'Testing a novel gesture-based mixing interface', *Journal of the Audio Engineering Society*, 61(5), pp. 301–313. Available at: www.aes.org/e-lib/browse.cfm?elib=16822 (Accessed: 6 January 2021).

Lee, S. and Kim, G. J. (2008) 'Effects of haptic feedback, stereoscopy, and image resolution on performance and presence in remote navigation', *International Journal of Human-Computer Studies*, 66(10), pp. 701–717. doi: https://doi.org/10.1016/j.ijhcs.2008.05.001

Liebman, N. *et al.* (2010) 'Cuebert: A new mixing board concept for musical theatre', in *Proceedings of the 2010 Conference on New Interfaces for Musical Expression. New Interfaces for Musical Expression (NIME 2010)*, Sydney, Australia, pp. 51–56. doi: 10.5281/zenodo.1177833

Mayer, R. E. and Moreno, R. (2003) 'Nine ways to reduce cognitive load in multimedia learning', *Educational Psychologist*, 38(1), pp. 43–52. doi: 10.1207/S15326985EP3801_6

Melchior, F. *et al.* (2013) 'On the use of a haptic feedback device for sound source control in spatial audio systems', in *Audio Engineering Society Convention 134*. Rome, Italy: Audio Engineering Society, p. 9. Available at: www.aes.org/e-lib/browse.cfm?elib=16742 (Accessed: 22 June 2022).

Mendelzon, A. O. (1996) 'Visualizing the World Wide Web', in *Proceedings of the Workshop on Advanced Visual Interfaces*. New York, NY: Association for Computing Machinery (AVI '96), pp. 13–19. doi: 10.1145/948449.948452

Miller, G. A. (1956) 'The magical number seven, plus or minus two: Some limits on our capacity for processing information.', *Psychological Review*, 63(2), p. 81. doi: 10.1037/h0043158

Mycroft, J. and Paterson, J. (2011) 'Activity flow in music equalization: The cognitive and creative implications of interface design', in *Audio Engineering Society Convention 130*. London, UK: Audio Engineering

Society, pp. 51–56. Available at: www.aes.org/e-lib/online/browse.cfm?elib=16568 (Accessed: 7 June 2022).

Mycroft, J., Reiss, J. D. and Stockman, T. (2013) 'The influence of graphical user interface design on critical listening skills', in *Proceedings of the Sound and Music Computing Conference 2013. SMC 2013*, Stockholm, Sweden, p. 5. doi: 10.5281/ZENODO.850419

Mycroft, J., Stockman, T. and Reiss, J. D. (2015) 'Audio mixing displays: The influence of overviews on information search and critical listening', in *International Symposium on Computer Music Modelling and Retrieval (CMMR). CMMR*, Plymouth, UK, p. 7. Available at: www.eecs.qmul.ac.uk/~josh/documents/2015/Mycroft%20Stockman%20Reiss%20-%20CMMR2015.pdf (Accessed: 16 August 2022).

Mycroft, J., Stockman, T. and Reiss, J. D. (2016) 'Visual information search in digital audio workstations', in *Audio Engineering Society Convention 140*. Paris, France: Audio Engineering Society, p. 8. Available at: www.eecs.qmul.ac.uk/~josh/documents/2016/Mycroft%20-%20AES140.pdf (Accessed: 12 July 2022).

Norman, D. A. (2002) The design of everyday things. Reprint. New York: Basic Books.

Øvergård, K. I. *et al.* (2007) 'Knobology in use: An experimental evaluation of ergonomics recommendations', *Ergonomics*, 50(5), pp. 694–705. doi: 10.1080/00140130601168046

Prewett, M. S. *et al.* (2006) 'The benefits of multimodal information: A meta-analysis comparing visual and visual-tactile feedback', in *Proceedings of the 8th International Conference on Multimodal Interfaces. ICMI '06*, Banff, Alberta, Canada: Association for Computing Machinery, NY, pp. 333–338. doi: 10.1145/1180995.1181057

Quiroz, D. and Martin, D. (2020) 'Exploratory research into the suitability of various 3D input devices for an immersive mixing task. Part II.', in *Audio Engineering Society Convention 149*. 10 pages. Online: Audio Engineering Society. Available at: www.aes.org/e-lib/browse.cfm?elib=20979

Ratcliffe, J. (2014) 'Hand motion-controlled audio mixing interface', in *Proceedings of New Interfaces for Musical Expression (NIME). 14th International Conference on New Interfaces for Musical Expression*, Goldsmiths, University of London, pp. 136–139. Available at: www.nime.org/proceedings/2014/nime2014_518.pdf (Accessed: 29 June 2022).

Robertson, G. G., Card, S. K. and Mackinlay, J. D. (1993) 'Information visualization using 3D interactive animation', *Association for Computing Machinery*, 36(4), pp. 57–71. doi: 10.1145/255950.153577

Scott, D. S. (2018) *The Way We Were: Mixers Past & Present (Part 1), Production Sound&Video*. Available at: www.local695.com/magazine/the-way-we-were-mixers-past-present-part-1/ (Accessed: 13 May 2023).

Shneiderman, B. *et al.* (2010) *Designing the User Interface: Strategies for Effective Human-Computer Interaction*. 6th Edition. Hoboken, NJ: Pearson.

Swedien, B. (2003) *Make Mine Music*. Norway: MIA Musikk.

Tan, Y. H. et al. (2015) 'Ergonomics aspects of knob designs: a literature review', Theoretical Issues in Ergonomics Science, 16(1), pp. 86–98. doi: 10.1080/1463922X.2014.880530

Théberge, P. (1997) *Any Sound You Can Imagine: Making Music/Consuming Technology*. Hanover, NH: Wesleyan University Press.

Wakefield, J., Dewey, C. and Gale, W. (2017) 'LAMI: A gesturally controlled three-dimensional stage Leap (Motion-based) Audio Mixing Interface', in *Audio Engineering Society Convention 142*. Berlin, Germany: Audio Engineering Society. 10 pages Available at: www.aes.org/e-lib/browse.cfm?elib=18661 (Accessed: 10 December 2022).

Whitenton, K. (2013) *Minimize Cognitive Load to Maximize Usability, Nielsen Norman Group*. Available at: www.nngroup.com/articles/minimize-cognitive-load/ (Accessed: 19 January 2023).

Wickens, C. D. (2002) 'Multiple resources and performance prediction', *Theoretical Issues in Ergonomics Science*, 3(2), pp. 159–177. doi: https://doi.org/10.1080/14639220210123806

Wolfe, J. M., Klempen, N. and Dahlen, K. (2000) 'Postattentive vision', *Journal of Experimental Psychology: Human Perception and Performance*, 26(2), pp. 693–716. doi: 10.1037/0096-1523.26.2.693

5

WHOSE D(ART)A IS IT ANYWAY?

Repositioning Data and Digital Ethics in Remote Music Collaboration Software

Martin K. Koszolko and Kristal Spreadborough

5.1 Introduction

Remote Music Collaboration Software (RMCS) platforms are used by musicians, music producers, and music production students and teachers, as they facilitate social networking, virtual studio collaboration, live music performances over the Internet as well as learning and knowledge transfer among the communities of users. RMCS has driven innovation in music practice through facilitating new forms of collaboration, engaging songwriters, performers, and music producers. However, as users engage with these platforms, they provide a wide variety of data, from personal information to artistic inputs to social networking details. This user information underpins the functionality of many social networks.

Findings and discussion presented in this chapter are a result of a pilot study in our ongoing research into RMCS. In this study, we have moved away from the impact of RMCS on music production, covered in our past research (see for example, Koszolko 2015, 2017, 2022), with the intention to highlight various issues connected to digital and data ethics. There is a growing public discourse about the appropriate use and storage of user-generated data by companies (Martin 2022, Shukla et al. 2022). This is often discussed in terms of data and digital ethics. The data and digital ethics of RMCS are yet to be problematised, despite the fact that these platforms touch upon a broad spectrum of pedagogical, social, artistic, and commercial activities for users. While RMCS represents significant opportunity for the music community, it is also important to understand how the data and digital practices that underpin the software's functionality are used and governed, and, by extension, how user rights are managed within this space.

In this study, we aim to address this gap in knowledge by examining selected aspects of the user experience and identifying related pressing digital and data ethics questions that are unique to RMCS. We focus on examples of three platforms from one subset of RMCS, which enables *Virtual Studio*-based music creation. Other types of RMCS, outside of our immediate focus here include *Telematic Platforms* and *Social Networking Platforms* (Koszolko 2024). The three platforms that we discuss in this chapter are BandLab, Endlesss and Soundtrap. They facilitate a comprehensive range of collaborative practices inclusive of synchronous and asynchronous remote music production, as well as social interactions and music distribution.

DOI: 10.4324/9781003396710-5

5.2 Rmcs Background

5.2.1 Evolution of the Software

While early remote music collaborations utilising computer networks can be traced as far back as 1978 and the activities of the League of Automatic Music Composers operating in the US, the most significant developments in this field took place in the last three decades and can be associated with the emergence of fast Internet networks, growing capacity of data storage, and more broadly, the development of Web 2.0. (Koszolko 2015, 2022). More recently, developments such as decentralised and token-based technologies are also being trialled in RMCS, of which one example is Endlesss (Endlesss Blog 2021).

In recent years, we have witnessed a range of shifts in the collaborative music software landscape. Some platforms have been acquired by larger music technology companies; for instance, Blend was acquired by Roli in 2015, and Soundtrap by Spotify in 2017. The case of Blend illustrates how such tech acquisitions can effectively halt software development, rendering an RMCS platform outdated and slowly abandoned by its user base. The story of Soundtrap's acquisition is more positive, as the platform has been actively developed under Spotify's ownership and, in 2023, was sold back to its founders, returning it to an independent operation (Smith 2023). Other examples, such as the closure of the Ohm Studio platform in 2021, demonstrate difficulties in monetising even very innovative and well-designed collaborative products. Alongside the emergence of newer platforms like Endlesss and the decline of established ones such as Ohm Studio, various music production and communication tools have been continuously improved across RMCS. The time of the Covid-19 pandemic and its associated limitations on face-to-face work acted as a catalyst in expediting these advancements across selected products.

Among the software discussed here, BandLab and Endlesss incorporate social networking and creative crowdsourcing features. These platforms offer musicians the ability not only to discover new creative partners but also to produce new music with them in synchronous or asynchronous ways. It is worth noting that Soundtrap used to include crowdsourcing and public music profile functionality as core features. However, these features were removed in 2020 during Spotify's ownership of the platform, a decision that was met with a negative response from its user base (Soundtrap Support 2020). While synchronous and asynchronous collaborations remain possible on Soundtrap, without the online community, they require sending invitations to users whose contact details are already known to us.

Moving beyond community features, each platform demonstrates unique strengths and focus areas. Endlesss emphasises a performative approach to recording and offers only basic mixing functionality. It is also well-suited to electronic music creation. In contrast, BandLab and Soundtrap provide a more traditional Digital Audio Workstation experience, featuring MIDI and audio tracks, as well as comprehensive mixing and processing capabilities. All three platforms are available as mobile apps, with BandLab and Soundtrap catering to both iOS and Android users, while Endlesss currently supports only the iOS platform. Additionally, Endlesss offers standalone and plugin functionality on Windows and macOS, while the other two RMCS options are accessible via web browsers and can be used on Windows, Mac, and Chrome operating systems. In terms of pricing, Endlesss provides its peer-to-peer music production services free of charge, whereas BandLab and Soundtrap operate on a freemium model, offering only basic functionality at no cost.

5.2.2 Users and Their Practices

Our observations on creative practices of RMCS users stem from the past fieldwork and research projects conducted by one of the authors. These include collaborations on over a dozen musical compositions with over 40 participants located in various geographical locations on three continents: Europe, North America and Australia (Koszolko 2015) as well as coordination of the international Collaborative Music Contest, an event which by now has attracted over two hundred participating musicians from around the world (Koszolko 2024). Collaborative, virtual studio practices within RMCS take various forms that include crowdsourcing, jamming and remixing (Koszolko 2015) as well as learning from the community of practice (Koszolko 2024). The engagement of users in crowdsourced projects is heavily dependent on building one's profile within the virtual community as well as skilful implementation of available communication tools (Koszolko and Montano 2016, Koszolko 2017, 2024).

The two years of the global Covid-19 pandemic accelerated the steady growth in the user base of RMCS technologies that could be witnessed over the past decade. During the pandemic, the distinction between digital and physical togetherness has evolved, further propelled by remote working arrangements. Affected groups included music hobbyists as well as professionals. Widespread physical isolation and remote work have led to virtual music collaboration becoming a necessity for music production teams but also for many music and sound production students and teachers across various levels of education. The pandemic years were a time of intense growth of selected collaborative platforms which enabled their ongoing technical development. A recent study by Knapp et al. (2023) "demonstrated a large increase in Soundtrap's user base beyond five standard deviations beginning in March 2020" (ibid.). Drawing on the dataset supplied by the company, the aforementioned research revealed that between its inception in 2011 and early January 2021, Soundtrap had been used by more than 10 million people. The BandLab platform witnessed possibly the biggest increase in user growth of all RMCS, reaching 30 million users in 2021 and 60 million users in 2023 (Ingham 2023). However, even prior to Covid-19, many online platforms provided valuable opportunities for individuals to learn music production techniques and software skills through participation in various communities of practice (Koszolko 2024). These communities, as described by Wenger-Trayner (2015), consist of groups of people who share a concern or passion for a specific domain, learning and improving together through regular interactions—a concept that also applies to RMCS.

As highlighted in previous research (Koszolko 2022), RMCS is utilised to a large degree by amateurs. However, platforms such as Endlesss are also being used by established artists such as Imogen Heap (Endlesss Blog 2023). Furthermore, as most RMCS users prefer to be represented by nicknames rather than their real names, it is often difficult to ascertain their level of experience and engagement in the industry. As evidenced by various submissions received for the Collaborative Music Contest in both 2021 and 2022, a significant portion of users fall within the category of young or underage individuals, although this demographic tends to vary depending on the specific RMCS platform. Additional fieldwork (Koszolko 2015, 2022) indicates that many RMCS users do not hold membership in a Performing Rights Organisation in their respective countries, and they often do not actively pursue the monetisation of their musical creations. Instead, their primary motivation lies in the joy of creating music or the artistic merit it offers, rather than a financial gain.

For users, practical engagement with RMCS typically commences with the sign-up process and agreement to a set of terms and policies. These documents serve as the means to obtain consent from individuals. All three platforms provide links to their Terms of Use as well as Privacy Policies in their website footers and sign-up pages.

5.3   Digital Ethics and RMCS

Digital and data ethics is a growing field of research globally. Interest in this area of work has been fuelled by high-profile cases of misuse of data and digital spaces. For example, Facebook's

Cambridge Analytica case (Hinds et al. 2020), exposure of discriminatory algorithms for making parole predictions, and the Australian Robodebt scandal (The Robodebt Royal Commission 2023) are just a few high-profile cases. Recent work has begun critiquing aspects of digital and data ethics across a range of areas, some examples include critiquing search engine results (Noble 2018), data analytics (D'Ignazio and Klein 2020), and critiquing technology through the arts (Crawford and Joler 2018). Today, driven in large part by the Covid-19 pandemic, an increasing number of individuals are engaging in digital environments. This surge in online engagement inevitably triggers a series of ethical questions related to digital practices and data usage, and the RMCS sector is not exempt from these concerns.

While we examine various facets of digital ethics within the realm of RMCS, it is essential to acknowledge that the issues we are about to address are not unique to this space. Nevertheless, there is a lack of work that critically examines these concerns in the context of peer-to-peer music production software. This gap in understanding is an impediment to fostering digital agency among musicians and music communities, both in the present and the future.

Digital and data ethics covers a broad range of topics and domains. Among these, three areas have garnered considerable attention recently, and these are also the focal points of our investigation. These areas include user autonomy, the capacity for informed consent, and the ramifications of data harvesting. We have chosen to explore these specific aspects in the context of RMCS for the following reasons:

- These areas have recently become the focal points of regional legislation aimed at regulating the data collected by companies about their users.
- There is a limited number of RMCS platforms available to users if specific functionalities are required. For instance, music educators may require the software to comply with legislation such as the Children's Online Privacy Protection Act or Family Educational Rights and Privacy Act in the U.S. Therefore, users do not always have a wide range of providers to select from, as they might when choosing email service suppliers, for example. This creates the potential for a vendor lock-in effect and can exacerbate certain digital ethics issues.
- There is a growing body of research in these three areas, providing a solid theoretical background for each of these aspects.

5.3.1 User Autonomy

In this paper, we take user autonomy to mean:

- The degree of control a user has over what data and information is collected.
- How easily they can extract and move their data.
- The structure of a company more broadly, as this might impact how users' data and information is shared and managed.

An increasing number of regulations are being introduced to enhance users' control and autonomy within digital spaces. Public discourse regarding third-party data collection is gaining momentum. Nonetheless, it remains uncertain to what extent users comprehend their ability to control the data collected about them and the extent to which they take action based on this information (Habib et al. 2020).

In our investigation of the three platforms, we assessed user autonomy by examining a specific set of questions. This is not to say that these are the only questions relevant to autonomy, but that they were the most pertinent to our study at this stage. We first examined the location of the parent company and corresponding data protection laws. Soundtrap and Endlesss are both based in very similar jurisdictions (England and Sweden) and need to comply with very similar regulations (see Table 5.1). These regulations are the GDPR (General Data Protection Regulation 2018) and the Data Protection Act (2018) which are quite comprehensive. BandLab is based in Singapore and the relevant regulations here (Personal Data Protection Act 2012–2020 Revised Edition) are less comprehensive than in the GDPR and Data Protection Act. Nonetheless, there are explicit regulations governing the treatment of user data in all three regions. However, of note here is that the kind of data that are the focus of all three regulations refers primarily to 'personal data'. Other types of data that users typically upload to RMCS platforms, such as recordings and images, may not fall under these regulations. As a result, it remains unclear to what extent users have control over this non-personal data.

One way in which this is clarified within RMCS is through statements of copyright and intellectual property. In all three platforms, users' data in the form of recordings and images is defined broadly as part of 'content'. All three include typical statements relating to granting the

TABLE 5.1 RMCS companies, laws and IP statements

	BandLab *www.bandlab.com*	**Endlesss** *https://endlesss.fm*	**Soundtrap** *www.soundtrap.com*
Location of Parent Company	Singapore	England	Sweden
Relevant Data Protection Laws	Personal Data Protection Act 2012	Data Protection Act 2018	GDPR 2018
Statements of Intellectual Property	". . . a worldwide, non-exclusive, royalty-free, sub-licensable and transferable licence to access, use, reproduce, distribute, publicly display, publicly perform, adapt, synchronise, prepare derivative works of, compile, make available and otherwise communicate Your Content to the public . . ." Clause 3 (BandLab Terms of Use 2023)	". . . non-exclusive right to use your Content . . . " ". . . The License applies worldwide and is royalty-free." ". . . waive any moral rights that may be vested in you as creator" Clause 3 (Endlesss Terms of Use 2023)	". . . worldwide, non-exclusive, transferable, sub-licensable, perpetual, irrevocable, royalty-free and fully paid up right to access and use Your Content" ". . . you initiate an automated process to transcode any audio Content and direct us to store Your Content on our servers . . ." Clause 10 (Soundtrap Terms of Use 2023)
Ownership Structure	Part of Caldecott Music Group, with expansions in media, production, and AI	A UK company Endlesss Ltd, with a connection to the Endlesss Foundation, registered in Switzerland	Single company since June 2023 (previously owned by Spotify)

platform non-exclusive, royalty-free usage of user content (see Table 5.1). There does not appear to be any explicit acknowledgement that users' recordings and images constitute data in their own right, except for a clause in Soundtrap's Terms of Use (2023) which mentions that when users upload audio to the platform, it triggers an automated process for converting and storing the user's content on the company's servers (see Table 5.1). This lack of clarity makes it challenging to assess the extent of control users have over the usage of their data by the platform.

The loose definition of data and the unclear framing of how recordings and images themselves constitute data may have implications for users' autonomy concerning the future use of their artistic artefacts. This concern is relevant because the ownership structure of various companies may result in content creators' data being used in ways that are not explicit or expected at the time of their engagement with the platform.

5.3.2 Informed Consent

There are a few models of consent which have been developed in a variety of fields including health sciences, research ethics, and social sciences. In this paper, we consider the concept of informed consent in relation to users' engagement with RMCS. The general requirements for informed consent, and the meaning we take in this paper, are that it is freely given, specific, unambiguous, and revocable (see, for example, Breen et al. 2020).

Our assessment of informed consent was guided by the content available in the Privacy Policy and Terms of Use, as well as the presentation of these documents. The questions that we looked at for assessment in this paper were the readability of the documents, what information is available to users and when, and the specificity of that information. We use two metrics when assessing readability: document length and document density. Table 5.2 provides an overview of the comparative lengths of the Terms of Use, Privacy Policy and Cookie Policy for each RMCS. Document metrics such as total word length, number of unique words, and vocabulary density are provided alongside these. BandLab and Soundtrap are of comparable length and with comparable density scores. Endlesss documents are not as long but are similarly dense. Of note is that BandLab and Soundtrap's Privacy Policies tend to be more readable than the Terms of Use, with fewer unique words compared to total words. However, Endlesss' Privacy Policy is not more readable than the Terms of Use despite being of comparable length. All platforms make these key documents available on their home pages and sign-up pages. Further to that, BandLab includes a dedicated page explaining their approach to copyright, which we featured in Table 5.2 as well.

Nonetheless, the lack of a clear differentiation between personal data, data in the form of uploaded recordings and images, and other types of collected data remains ambiguous. As a result, it becomes challenging to ascertain the extent to which users can provide precise consent. Similarly, all three platforms contain disclaimers indicating that once data is shared with other arms of the company, utilised by other users, or transferred to other vendors, the platform cannot retrieve or delete that data. In this context, a user's consent is also not revokable.

5.3.3 Impacts of Data Harvesting

Narratives like 'data is the new oil', 'data is gold' and, quoting Shukla et al. (2022: v), "Data is the new Soil, Let us cultivate it", speak to the lucrative nature of the data market. This is exemplified in the increasingly high sale prices of social media tech companies (see, for example, Short and Todd 2017). One profitable avenue for data utilisation is through a process known as 'secondary use', which involves employing data for purposes beyond its original collection intent.

TABLE 5.2 Descriptive statistics concerning terms of use, privacy policy, copyright and cookies policy of the 3 discussed platforms. These statistics were derived using Voyant – an online NLP tool. All links were accessed in August 2023

Site	Link	Descriptive Statistics
BandLab Terms of Use (2023)	https://blog.bandlab.com/terms-of-use/	Document word length – 8948 words Vocabulary density – 0.172 Unique words – 1536
BandLab Privacy Policy (2023)	https://blog.bandlab.com/privacy-policy/	Document word length – 3252 Vocabulary density – 0.235 Unique words – 763
BandLab Copyright (2023)	https://blog.bandlab.com/learn-about-copyright/#copyrights	Document word length – 2344 Vocabulary density – 0.253
BandLab Cookie Policy (2023)	https://blog.bandlab.com/cookies-policy/	Document word length – 1017 Vocabulary density – 0.342 Unique words – 348
Soundtrap Terms of Use (2023)	www.soundtrap.com/legal/terms/creator/world	Document word length – 9500 Vocabulary density – 0.162 Unique words – 1540
Soundtrap Privacy Policy (2023)	www.soundtrap.com/legal/privacy/world	Document word length – 3396 Vocabulary density – 0.207 Unique words – 704
Soundtrap Cookie Policy (2023)	www.soundtrap.com/legal/cookies	Document word length – 938 Vocabulary density – 0.321 Unique words – 301
Endlesss Terms of Use (2023)	https://endlesss.fm/terms	Document word length – 3348 Vocabulary density – 0.238 Unique words – 796
Endlesss Privacy Policy (2023)	https://endlesss.fm/privacy	Document word length – 2892 Vocabulary density – 0.220 Unique words – 635
Endlesss Cookie Policy (2023)	https://endlesss.fm/cookies	Document word length – 677 Vocabulary density – 0.356 Unique words – 241

Our primary focus revolved around examining the commodification of users' data by scrutinising platform statements regarding the secondary use of this data. In all three cases, it remains unclear how precisely data will be employed in the future. Each platform contains statements indicating that it is not possible to guarantee that all of a user's data or content will be removed upon request. Each also indicates that cookie data may be shared with third parties. BandLab and Soundtrap provide information on how data will be shared both within and beyond the company, however, Endlesss's documentation is less detailed. Notably, none of the platforms specify what the data will not be used for. In summary, based on the information available to users in the Terms of Use, it is not possible to assess the potential secondary uses to which data may be subjected.

5.4 Findings and Future Work

RMCS systems serve not only as spaces for creative crowdsourcing and collaboration but also as valuable avenues for skill acquisition, whether through formal educational channels or more informal means. Moreover, these platforms facilitate the entire process of music creation, allowing individuals to compose, write, and record their music. Implementing RMCS in collaborative music practices constitutes a novel form of digital social practice (Haslanger 2018) between collaborating musicians, companies, and technologies. And through exploring the affordances and ethics of this practice, our aim is to foster informed engagement among musicians in the context of digital innovation.

The trends identified in the three platforms assessed in this project are not unique to RMCS. Rather, these are areas that are widely discussed within digital and data ethics communities. Collaborative software has played and will continue to play a vital role in musicians' communal practices. Our objective in this paper is to identify areas where we can enhance users' understanding of their digital agency. To this end, we identify three avenues of future work.

First, future work should examine how RMCS might provide more explicit definitions of data, including how artistic outputs are treated as data. While personal data are well defined, and data from cookies are often explicitly described, it is not clear how other 'content', such as audio and images uploaded and created by users, qualifies as data. These artifacts not only contain personal information from users, such as their voice, but also constitute their artistic outputs and intellectual property. Valuable insights can be drawn from the field of digital cultural heritage, which frequently deals with concepts of intellectual property and ownership of digitised physical items.

Second, future research should examine how documentation, such as Privacy Policies and Terms of Use, can be made more accessible to lay audiences. This is especially important in the case of RMCS, where users may have limited platform choices to engage and create music with the community. Work in the legal field has already begun to examine the readability of such documents from the perspective of consumer law. This includes addressing contract law doctrines like 'the duty to read' contracts, which can sometimes be quite challenging to decipher (Benoliel and Becher 2019). Guidance on how to enhance the approachability and readability of these documents for users may be drawn from this work.

Third, future work should explore how secondary use of users' data can be more explicitly communicated and any changes continuously conveyed. This is especially crucial for musicians creating content, as it may take time for their creations to yield economic benefits. How their data and content are used in the interim can have an impact on this process. Models for effective communication can be drawn from data breach reporting policies commonly used by IT companies. While secondary use of data differs from a data breach, data breach statements often need to convey complex technical concepts to non-technical audiences. Insights on how to translate this information effectively can inform the development of more explicit statements regarding the secondary use of data.

Furthermore, it would be advantageous to incorporate stakeholder engagement, involving software users and industry representatives, through discussions and interviews. These conversations could serve as valuable sources of input, enriching our understanding of the observed issues, and informing the strategies we formulate in response to the avenues of future work outlined above.

5.5 Conclusion

The process of datafication in the realm of music streaming, facilitated by Digital Service Providers, has been recognised as an issue that transforms the act of music consumption into a commodified dataset (Schwarz and Johansson 2022). This research underscores a comparable concern within the domain of peer-to-peer music composition and production.

Our exploration of several aspects of digital ethics in relation to a sample of collaborative music software highlights the importance of ethical awareness in innovation. As RMCS continues to shape music creation within networked spaces, cooperation among artists, platforms, and policymakers becomes essential for creating an environment that blends creative pursuits with respect for digital rights. By highlighting the potential ethical challenges and opportunities surrounding the use of RMCS, we can encourage the development of best practices that safeguard the rights of music creators, promote informed engagement, and foster ongoing innovation within the music industries. This considered approach to addressing the issues and questions raised in this chapter can lead to a repositioning of how user data is managed and communicated, ultimately benefiting the vast community of musicians who utilise these platforms.

References

BandLab Terms of Use. (2023). *BandLab Blog* (website), available online at: https://blog.bandlab.com/terms-of-use/ [accessed September 2023].

Benoliel, U. and Becher, S. I. (2019). *The Duty to Read the Unreadable.* 60 Boston College Law Review 2255, available online at: https://ssrn.com/abstract=3313837 [accessed August 2023].

Breen, S., Ouazzane, K., and Patel, P. (2020). GDPR: Is your consent valid?. *Business Information Review,* 37(1), 19–24.

Crawford, K., and Joler, V. (2018). Anatomy of an AI System: The Amazon Echo As An Anatomical Map of Human Labor, Data and Planetary Resources (website) available online at: https://anatomyof.ai [accessed August 2023].

Data Protection Act. (2018). *Gov.uk* (website), available online at: www.gov.uk/data-protection [accessed September 2023].

D'Ignazio, C., and Klein, L. (2020). Acknowledgments, *Data Feminism,* available online at: https://data-feminism.mitpress.mit.edu/pub/f8vw7hhr [accessed July 2023].

Endlesss Blog. (2021). Endlesss Transition Statement, *Endlesss* (website), available online at: www.blog.endlesss.fm/post/endlesss-transition-statement [accessed August 2023].

Endlesss Blog. (2023). The Endlesss Beat Machine: The Genesis of New Music, *Endlesss* (website), available online at: www.blog.endlesss.fm/post/the-endlesss-beat-machine-the-genesis-of-new-music [accessed August 2023].

Endlesss Terms of Use. (2023). *Endlesss* (website), available online at: https://endlesss.fm/terms [accessed September 2023].

General Data Protection Regulation. (2018). Intersoft Consulting (website), available online at: https://gdpr-info.eu/ [accessed August 2023].

Habib, H., Pearman, S., Wang, J., Zou, Y., Acquisti, A., Cranor, L. F., Sadeh, N., and Schaub, F. (2020). "It's a scavenger hunt": Usability of Websites' Opt-Out and Data Deletion Choices. *Proceedings of the 2020 CHI Conference on Human Factors in Computing Systems,* pp. 1–12.

Haslanger, S. (2018). What is a social practice? *Royal Institute of Philosophy Supplement,* 82, 231–247.

Hinds, J., Williams, E. J., and Joinson, A. N. (2020). "It wouldn't happen to me": Privacy concerns and perspectives following the Cambridge Analytica scandal. *International Journal of Human-Computer Studies,* 143

Ingham, T. (2023). Spotify Wants 50 Million Creators on Its Platform by 2030. Bandlab Already Has 60 Million, (website), available online at: www.musicbusinessworldwide.com/spotify-wants-50-million-creators-bandlab-already-has-60-million/ [accessed July 2023].

Knapp, D. H., Powell, B., Smith, G. D., Coggiola, J. C., and Kelsey, M. (2023). Soundtrap usage during COVID-19: A machine-learning approach to assess the effects of the pandemic on online music learning. *Research Studies in Music Education,* 45(3)), 571–584.

Koszolko, M. K. (2015). Crowdsourcing, jamming and remixing: A qualitative study of contemporary music production practices in the cloud, *Journal on the Art of Record Production,* 10. www.arpjournal.com/asarpwp/crowdsourcing-jamming-and-remixing-a-qualitative-study-of-contemporary-music-production-practices-in-the-cloud/

Koszolko, M. K. (2017). The giver: A case study of the impact of remote music collaboration software on music production process, *IASPM Journal*, 7(2), 32–40.

Koszolko, M. K. (2022). The Virtual Studio, in G. Stahl and J. M. Percival (eds.) *The Bloomsbury Handbook of Popular Music, Space and Place*. New York: Bloomsbury Academic, pp. 217–228.

Koszolko, M. K. (2024). Connecting Across Borders: Communication Tools and Group Practices of Remote Music Collaborators, in J-O. Gullö, R. Hepworth-Sawyer, J. Paterson, R. Toulson, and M. Marrington (eds.) *Innovation in Music: Cultures & Contexts*. London: Routledge, pp. 247–263.

Koszolko, M. K., and Montano, E. (2016). Cloud connectivity and contemporary electronic dance music production, *Kinephanos*, *Journal of Media Studies and Popular Culture*, 6(1), 60–86.

Martin, K. (2022). *Ethics of Data and Analytics: Concepts and Cases*. Boca Raton: CRC Press.

Noble, S. U. (2018). *Algorithms of Oppression: How Search Engines Reinforce Racism*. New York, New York: New York University Press.

Personal Data Protection Act 2012–2020 Revised Edition. (2021). *Singapore Statutes Online* (website), available online at: https://sso.agc.gov.sg/Act/PDPA2012 [accessed September 2023].

Schwarz, J. A., and Johansson, S. (2022). When music becomes datafied: Streaming services and the case of Spotify, in S. Homan (ed.) *The Bloomsbury Handbook of Popular Music Policy*. New York: Bloomsbury Academic, pp. 289–304.

Short, J. E., and Todd, S. (2017). What's your data worth? *MIT Sloan Management Review*, 58(3), 17–19.

Shukla, S., George, J. P., Tiwari, K., and Kureethara, J. V. (2022). *Data Ethics and Challenges*. Singapore: Springer.

Smith, D. (2023). *Almost Six Years Later, Spotify is Selling Soundtrap Back to Its Founders*, available online at: www.digitalmusicnews.com/2023/06/12/soundtrap-spotify-sale/ [accessed August 2023]

Soundtrap Support. (2020). (Website), https://support.soundtrap.com/hc/en-us/community/posts/360010156 919-Do-Not-Remove-Public-Profiles [accessed August 2023].

Soundtrap Terms of Use. (2023). *Soundtrap* (website), available online at: www.soundtrap.com/legal/terms/creator/world [accessed September 2023].

The Robodebt Royal Commission. (2023). *Royal Commission into the Robodebt Scheme* (website), available online at: https://robodebt.royalcommission.gov.au/ [accessed August 2023].

Wenger-Trayner, E., and Wenger-Trayner, B. (2015). *Communities of Practice a Brief Introduction*. Available at: http://wenger-trayner.com/introduction-to-communities-of-practice/ [accessed July 2023].

6

LISTENING AS CONTEMPLATION

A Reflexive Thematic Analysis of Listening to Modular-based Compositions

Rotem Haguel and Justin Paterson

6.1 Introduction

> Thus, he develops his awareness to such an extent that there is *mere understanding* along with *mere awareness*.
>
> (Mahasatipatana Sutta, Goenka, 1993, p. 29)

My (Haguel's) first meditation retreat sparked an interest in studying the mind, a field characterised by a gap between the physical brain and the vast range of human experiences (Clarke and Clarke, 2011). This gap, known as the 'hard problem' of consciousness (Montague, 2011), cannot be fully resolved through empirical science, as it requires subjective, first-person data. Meditation, according to some (Lowe, 2011; Varela, Thompson and Rosch, 2016) is often considered an alternative approach to studying the mind. My experience working with hardware synthesisers further deepened my exploration of mind and the nature of moment-to-moment experience. Manipulating synthesisers taught me the importance of attentive listening, as small gestures could produce significant changes in sound.

While existing scholarly literature focuses on the design, components, and creative possibilities of modular synthesisers (Bruno, 2016; Hyde, 2016; Navs, 2016; Hetrick, 2017; O'Connor, 2019), it rarely examines the experience of listening and practicing with these instruments. Inspired by these insights, I aim to investigate how compositions created with modular synthesisers can reflect contemplative insights. By 'contemplative', I refer to insights gained from observing physical and mental phenomena through meditation. Although there are various meditation practices and experiences, I will draw from my own knowledge practicing vipassana meditation to explore this unique intersection.

6.2 Background

Vipassana (meaning 'insight', in Pali) is a practice taught by Burmese/Indian teacher S. N. Goenka. It involves developing awareness of the changing nature of physical and mental experience. The teaching is centred around three techniques—*ānāpānasati*, *vipassana* and *metta bhavavna*. During practice, students gain certain insights, two of which are most relevant here: 1) *impermanence*

DOI: 10.4324/9781003396710-6

is appreciated through the experience of transitory sensations; 2) the notion of *non-self* becomes apparent when one considers the flux of psychosomatic phenomena (Hart, 2018, pp. 94–95). Understanding these realities begins with ānāpānasati, which involves mindfulness or attention of one's inhalations and exhalations.

Attention is defined as a cognitive process essential for the selection and processing of information from the surrounding environment, with a controlled and limited capacity (Styles, 2006). While the nature of control remains debatable, attention can certainly be conceived as finite and can, to some extent, be directed. Attention of auditory information is often described as listening, thus composer and performer, Pauline Oliveros (2005) distinguishes *hearing* as an involuntary, physical process from *listening*, which involves attention. For artist-researcher, Biswas (2011, p. 103), listening also entails "restraint from habitual, involuntary reactions", which establishes it not just as an attentive act, but as one that involves concentration.

Once concentration is established, students are introduced to the technique of vipassana, which involves systematically scanning one's body and observing sensations non-verbally and without judgment. In this form of examination, practitioners realise that every sensation that arises, whether pleasant, unpleasant, or neutral, eventually fades away. Directly experiencing these fluctuations, practitioners learn about the reality of impermanence, the cultural significance of which is discussed by Geismar, Otto, and Warner (2022). A universal condition that stands in opposition to a human desire for fixity, impermanence is central to various aspects of our lives, including our aging bodies, the transient nature of material possessions, shifts in social status and cultural practices, and environmental changes.

Yet insight into impermanence goes beyond our common understanding of "the leaves-fall, maidens-wither, and kings-are-forgotten", as neurobiologists Varela, Thompson, and Rosch (2016, pp. 60–61) emphasise. Rather, it relates to the "personal penetrating impermanence of the activity of the mind itself [. . . as a] shifting stream of momentary mental occurrences". Music serves as a platform for embodying this unique perspective. Not only does it convey flow and change through its physical manifestations, but it also enables individuals to experience transformations in their emotional and physical states (Biswas, 2011).

Since all aspects of our mental and physical structure arise and cease, none can be identified as 'self' in the stable and enduring sense of the word. The flux of an individual's experience of themselves and the world underpins the Buddhist teaching on non-self (Goenka, 2017). This notion is contained within the idea of emptiness – a key principle in Mahāyāna Buddhism which holds that all things lack inherent existence, as they arise from causes and conditions and exist within a network of relationships (Priest, 2009). Music, according to Lowe (2011, p. 119), offers a suitable contemplative medium for emptiness. She considers the sound of an instrument or its pitch to be "gross simplifications", empty of "self-existence", since their spectrographic image reveals them as "fluxing, inter-dependent, baseless phenomena".

Musical experiences and meditative states often exhibit transcendent qualities, that is, they surpass or exceed ordinary experience. Wahbeh et al. (2018) define transcendence affiliated with meditation as an indescribable, subjective experience of an altered state of consciousness, whereas Aldridge (2003) describes it as a leap beyond current awareness to a deeper level of understanding. Several authors have explored the role of music in facilitating transcendent experiences. Gabrielsson's *Strong Experiences with Music* (2011), offers a comprehensive and systematic examination of music's transformative power. Drawing on two decades of research, Gabrielsson dedicates a chapter to 'Music and Transcendence', which highlights experiences described as magical, mysterious, supernatural, or even extra-terrestrial. In transcendent experiences, individuals may feel as if they are in a trance or ecstasy, experiencing a complete merging with something greater and catching glimpses of other realms of existence.

In her book *Deep Listeners* (2004), Becker examines transcendent experiences facilitated by music within ritualistic and religious contexts. She differentiates between meditation, conducted in solitude, stillness, and silence, and trancing, which occurs in a communal setting accompanied by music and sensory stimulation. Interestingly, both experiences of trancing and meditation involve a diminished sense of self or a temporary replacement of the autobiographical narrative we construct about ourselves.

In a similar vein, Clarke (2014) traces a theme in Western thought, according to which music allows individuals to both 'lose' and 'find' themselves within it. Clarke argues that these states are possible when our linguistic, *higher-order* consciousness becomes quiet, revealing a pre- or non-linguistic *primary* consciousness. He acknowledges the difficulty of quieting higher-order consciousness but asserts that significant glimpses of primary consciousness can be regained through meditative practices and sensorimotor experiences, including music.

In the Mahāsatipatthāna Sutta, the Buddha identifies the goal for practicing mindfulness: 'mere awareness', a state transcending conventional subject-object conceptualisation, wherein there is no 'I' looking over at experience. Wahbeh et al. (2018) explore this nondual aspect of transcendent experiences. They define nonduality as a state of 'background awareness' that goes beyond conceptual cognition, knowing itself to be conscious reflexively. Nonduality is described as a unifying space that presupposes the contents of consciousness, devoid of such dualistic fragmentation as 'this or that' or 'then and now'.

6.3 Methodology

In examining how meditation can inform listening to works made with modular synthesisers, I assume an interpretivist paradigm, studying subjective experience from within (Cohen, Manion and Morrison, 2017). Furthermore, in examining my own experience, my research is informed by principles of self-study, framed by autoethnography. Autoethnography involves writing and systematically analysing personal experience as a means of gaining cultural insight (Ellis, Adams and Bochner, 2011). Departing from an observation of self (auto), it aims at revealing cultural understandings (ethno) through evocative descriptions (graphy) (Adams, Holman Jones and Ellis, 2021). In autoethnographic works, the researcher becomes the researched and, through a process of recalling and self-observation, seeks rich narratives with a cultural interpretive orientation (Chang, 2008). This framework enables me to study listening through the lens of my own background in meditation and electronic music-making.

My approach is also informed by the practice of deep listening (Oliveros, 2005), which involves expanding one's auditory awareness through an attendance to the spectrum of sound/ silence. A form of sonic meditation, deep listening emphasises attaining knowledge through an embodied engagement with the world. Embodied approaches to listening implicate the body in musical perception (Borgo, 2012), and in the context of cognition, they emphasise a reflection that bridges mindfulness, phenomenology and cognitive science (Varela, Thompson and Rosch, 2016). While deep listening offers an embodied approach for attending to sound, its interface with autoethnography provides a systematic and structured approach for investigating modular-based works.

6.4 Procedure

My approach utilises systematic self-observation, an autoethnographic method for tracking researchers' behaviour during activities of interest (Chang, 2008). The same logic is expressed by

Oliveros (2005, p. 17), who encourages practitioners to journal their experiences, with reference to aspects like "dynamics, feelings, sounds and sensations". Through journaling, my hope was to learn what other composers *do* when they make 'contemplative' works and what it is like listening to their work. Such questions, I believed, would reflect my interest in the qualia of experience, as well as the practicalities of modular-based composition.

In preparation for listening, I compiled 16 electroacoustic pieces that are affiliated with contemplative insights and modular-based practices. For example, Èliane Radigue's piece 'Kyema' (1994), draws inspiration from Tibetan Buddhism and was created using an ARP modular system (Radigue, 2009; Wu, 2020). Similarly, Caterina Barbieri's works utilise a bespoke Eurorack system, exploring themes like consciousness, listening, and memory (2017; 2018). Between July and August 2022, I conducted 16 listening sessions and documented my experiences in journals. A list of journals and their (subsequently referenced) abbreviations can be found in Table 6.1 below.

The journals were analysed using Braun and Clarke's (2022) Reflexive Thematic Analysis (Reflexive TA) approach. Thematic analysis is a research method that aims to identify patterns and meaning in data across cases. Its reflexive variety differs in that it situates itself within a qualitative paradigm, embracing the inescapable subjectivity of data interpretation. Thus, Reflexive TA encourages critical reflection on the researcher's role and the research practice. Incorporating Reflexive TA within this research enables me to embrace my position as a researcher and the way it shapes my meaning-making process. In addition, it allows to approach this process with an open question and through data generation and contextualisation, to adapt and refine that question accordingly.

I began familiarising myself with the data by transcribing hand-written journal entries into rich text files and importing them into NVivo software. This helped identify repeated words, phrases, and concepts, which formed the basis of two coding cycles. After the second cycle, initial themes emerged and were further developed, refined, and named. For many people, this process is often messy, as each stage, including the write-up, informs the themes and their labels. The map below (Figure 6.1) shows the generated themes, subthemes, and their relationship to one another. In the following sections, I will attempt to describe each theme, the different facets or codes organised

TABLE 6.1 List of journal abbreviations

Code	Artist	Piece	Length	Date
BU	Abul Mogard	Bound Universe	00:06:45	01/08/2022
MUT	Barker	Maximum Utility	00:07:26	29/07/2022
UT	Barker	Utility	00:04:30	31/08/2022
PNU	Caterina Barbieri	Pnuema	00:11:01	10/08/2022
TCCTF	Caterina Barbieri	This Causes Consciousness to Fracture	00:14:45	08/08/2022
INTCAEB	Caterina Barbieri	INTCAEB	00:09:22	11/08/2022
SOTRS	Caterina Barbieri	SOTRS	00:09:30	10/08/2022
TLL	Caterina Barbieri	The Landscape Listens	00:08:17	11/08/2022
VW01	Donatto Dozzi	Vaporware 01	00:05:49	10/08/2022
KYE	Èliane Radigue	Kyema	01:01:22	13/08/2022
DA	JakoJako	Deine Augen	00:05:17	08/08/2022
MPVV	Jonathan Harvey	Mortuos Plango, Vivos Voce	00:09:08	02/08/2022
XE	Kara-Lis Coverdale	X 4Ewi	00:05:20	10/08/2022
DI	Lisa Bella Donna	Double-Image	00:12:49	04/08/2022
DPV3	Venetian Snares	Dreamt Person v3	00:01:56	10/08/2022
LAWL	Abul Mogard	Live @ Waking Life	01:00:00	29/08/2022

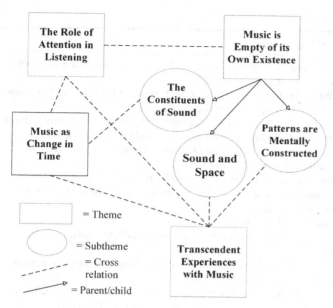

FIGURE 6.1 Thematic map.

within it, and how they manifest in the data. In so doing, my hope is to draw a boundary around each theme, all the while exploring its relationships with the rest.

6.5 Theme 1: The Role Of Attention In Listening

The term 'attention' appears frequently throughout the dataset, as the journals explore various aspects such as shifts of attention, types of attention, and concentration. Some data segments even highlight the experience of a present-moment temporality in attention and the reflexive capacity it enables. Generated at an early stage of analysis, theme 1 was labelled '*the role of attention in listening*'. It conceptualises listening as an act of directing attention to auditory information and emphasises the structure and form of attention to sound. The codes belonging to this theme are seen in Table 6.2, alongside examples of coded segments. The following paragraphs will provide a description of each aspect belonging to this theme and the way it pertains to the data.

Borrowing from Styles (2006), the journals provide multiple examples for the directedness and finitude of attention. The excerpt from UT, for example, demonstrates the inextricable link between the directed and limited capacity of attention. When we listen, we experience 'shifts' not only between different objects within the auditory scene but also between different auditory perspectives. This relates to a distinction made by Oliveros, between 'global' and 'focal' attention, as she writes:

> Focal attention, like a lens, produces clear detail limited to the object of attention. Global attention is diffused and continually expanding to take in the whole of the space/time continuum of sound. Sensitivity is to the flow of sound and details are not necessarily clear. (2005, p. 13)

These shifts occur exactly because of the bandwidth of attention, its finite nature, and this is further illustrated in instances where the mind 'wanders away' or when concentration is 'lost'. When we

TABLE 6.2 Coded segments for theme 1

Code	Example Journal Extracts
Attention can be directed	"At points, my listening becomes *global*, and I hear the sounds, their echoes, and their reverberance all at once" (TCCTF)
	"I noticed my attention shifted from moments where [. . .] I was focused on a single sound and the way all sounds appear together" (UT)
	"Indiscernible at first, [the tone] quickly becomes more present. My focus is drawn to its shifting spectral quality, its inner dynamics and timbre" (KYE)
Attention is finite	"I found myself losing concentration, and when I came back, I noticed certain elements are different now [sic]" (DA)
	"There was a moment where my mind wandered" (BU)
	"I found myself fighting the urge to look at my phone, check the time, go to the toilet, and not always succeeding" (KYE)
Attention affords reflexivity	"I noticed my attention [has] shifted" (MUT)
	"[This] reminded me of the sound of my own being. A reflexive awareness in listening" (BU)
	"The disappearance of sound meant that I could momentarily witness awareness itself, and that was very powerful" (LAWL)
Attention in the present moment	"The steady growth of the piece, its tempo and moving elements [. . .] facilitate a moment-to-moment observation of unfolding phenomena" (SOTRS)
	"In moment-to-moment experience my consciousness observes what appears to be fixed objects" (YE)
	"Sound as a phenomenon comprises endless changes in each moment, all framed through our memory and anticipation" (MPVV)

lose concentration, we are subject to an involuntary shift of attention. Indeed, this unintentional drift challenges the controlled nature of attention, for it shows two sides to the direction of attention: the voluntary vs. involuntary. Concentration as an act of sustained attention is seen across the data set, and journals excerpts, like KYE, often show effort to maintain concentration despite habituated patterns and interruptions. This evokes the technique of ānāpānasati, where students quickly recognise the ease with which their mind wanders away from the breath to a thought or another sensory experience. Segments pointing toward a tension between concentration and distraction ('phone', 'time' and 'toilet') imply something further—like in ānāpānasati, attention can be trained.

Another aspect of this theme relates to the temporality of attention, as some segments refer to attention to moment-to-moment experience of sound. The observation of experience from moment to moment connects the listener to a phenomenological present charged with the memory of the past and an anticipation towards the future. Paying attention enables us to recognise change in what would otherwise appear fixed, and this will be further explored in the discussion below. The last aspect refers to how attention might be turned back at itself, enabling a particular type of reflexivity. Reflexivity is implicit in the very act of journal-writing and conducting research, yet here we see a reflexivity in attention itself—attention of attention.

> LAWL: The disappearance of sound meant that I could momentarily witness awareness itself, and that was very powerful.

This idea of an awareness underlying all phenomena can be described in terms of nonduality (Wahbeh *et al*., 2018), and this relates to the theme on transcendence, discussed below.

6.6   Theme 2: Music As Change In Time

Formed around notions of musical process, materiality, and musical time, this theme examines music as a vessel for change. It recognises the mechanisms by which change (material or compositional) occurs and is perceived, while reflecting an understanding that, as a matter of experience, an awareness of change is synonymous with or, at the very least, linked to our perception of time. Table 6.3 provides an overview of codes and subcodes constituting this theme, including example segments for each.

The first aspect, '*Music as process*' borrows from the work of minimalist composer, Steve Reich (2017), who famously described his attempt to lay bare a slow musical process and watch

TABLE 6.3 Coded segments for theme 2

Code	Subcode	Example journal extracts
Music as process	Type of process	"The piece opens with a [. . .] a sparse pattern, and, very clearly, I hear it undergoing an additive process, whereby notes are added at the end of each pattern cycle." (TCCTF)
		"There is a strong sense of pulse throughout until the sequence starts to speed up, and this process ends when it almost multiplies its speed." (DPV3)
		"The music was predictable in its progression, yet at the same time slow and gradual, [which] meant you were transfixed by its movement" (LAWL)
	Process needs to be audible	"It's as if notes were inserted into a system with rules I cannot hack, but I know there are rules nonetheless and they are somewhat observable." (VW01)
		"While [. . .] the piece utilises modular synthesis [. . .] it doesn't bring to the fore any process, and so I am left wondering – what do I take from this?" (DA)
Change to sonic material	Material transformation	The piece evokes a sense of materiality in me [. . .] It does this in a myriad of ways: 'real' sounds (bells) are played as they are; they are reversed; their envelope is altered; they are constructed using additive methods (synthesis); voices are blended with the tones, fragmented to create pops, and clicks. (MPVV)
		"As repetition of musical material ensues, I focus more on the timbral transformation of sound" (MU)
	Modulation	"When the envelope became sharp, the bursts would dance around the stereo field." (SOTRS)
		"At first, I hear an oscillating tone and my attention focuses on it" (PNU)
		"It rises like a wave and recedes" (XE)
Change is linked to a sense of time	A sense of meter	"At first, when the sequence starts, the serenity of the system strikes me. It is slow and even. No odd time, very repetitive." (INTCAEB)
		"The piece exhibits a hypnotic quality. Perhaps it is the combination of the tempo, and incessant repetition, followed by a tide-like process, 'washing' over." (SOTRS)
		"Sounds [are] 'smeared' by the sound of reversed reverbs. This causes a feeling of a-temporality." (TLL)
	The role of percussive elements	The result [. . .] is punctuated by bursts of percussive energy at different moment of the sequence. (DPV3)
		"The track starts with percussive noises that establish a sense of meter. It feels like there is an implied beat and I feel like moving. It is relatively up-tempo." (UT)

it unfold. Thus, journal segments were assigned with this code, as they referred to different types of processes applied to pitch, note length, and tempo. Whether deterministic (e.g., linear, or additive) or random, the processes had to be *heard* and this is seen in VM01 and DA. One form of processing—reversal—is mentioned in several journals, and in one case it is described in terms of 'smearing' (TLL). Of course, this image evokes the metaphor of sound as material, which can be situated within debates on the aesthetics of electronic music. In her book *Listening Through the Noise*, Demers (2010) argues that sound in electronica is often portrayed as being malleable, and her taxonomy of three 'activities' in electronic music (construction, reproduction and destruction) builds on that theme. In that vein, segments in MPVV and SU journals were assigned with the code '*Change to sonic material*' since they utilise the same metaphor.

The last aspect labelled '*Change is linked to a sense of time*' is concerned with notions of musical time, its perception, and punctuation. As such, it includes segments that refer to the tempo and meter of a piece and the feeling it invokes—be it calmness, stability, urgency, or even disorientation. It also refers to a sense of 'pulse' or 'beat' and includes instances where this is accentuated by virtue of percussive elements with strong transients. Change forms our perception of time, and music can facilitate or alter our experience of time through its tempo, meter, and rhythm.

Notions of process, temporality and material transformation are interpreted through the idea of impermanence, as seen in the following excerpt:

> KYE: Occasionally, I notice great change has occurred to the soundscape, one that occurred without me noticing. This is a huge lesson in impermanence [. . .] But it is also a lesson in memory and our clinging to what occurred in the past.

Indeed, sound is a vessel for change, but real insight lies in appreciating subtle change in all things. While listening, we might note latent, imperceptible change through temporal breaks of continuity.

6.7 Theme 3: Music Is Empty Of Its Own Existence

An understanding of sound as it communicates impermanence challenges our perceptions of fixed, stable objects. When one comes close to what is truly happening, one finds 'cracks' in concepts that would otherwise seem solid and perpetual. This underlying notion provides context for another latent theme generated in the dataset, one which relates to the insubstantiality of sound, its empty nature. The label '*Music is empty of its own existence*' was assigned to segments that reveal music's lack of 'self-existence'. Comprising three separate subthemes, this theme poses questions like 'What constitutes a sound?', 'What constitutes a musical pattern?' and 'What is the relationship between sound, space and silence?'. These questions correspond with the subthemes below, seen in Table 6.4.

6.7.1 Subtheme 3.1: The Constituents of Sound

> MUT: I start noticing the different constituents of the piece, picking them apart as well as appreciating them together.

Sounds in a musical piece are often conceived as of being unified, yet upon closer examination, we might notice their constituent elements. The term 'constituents of sound' refers to the sonic components or aspects of sounds within the composition. It focuses on auditory qualities rather than on conceptual or intellectual aspects. Harvey's composition 'Mortuos Plango, Vivo Voce'

TABLE 6.4 Coded segments for theme 3

Subtheme	Code	Example Journal Extracts
The constituents of sound	The bell	"I was drawn to the timbral quality of the real bells and bell-like tones [. . .] It reminds me of the gong in meditation." (MPVV)
	Additive/ granular synthesis	"Fragmented to create pops and clicks." (MPVV)
	Timbre and overtones	"The tone, while singular, is accompanied by discernible frequencies in consonance" (KYE)
A sense of space	Nondual aspects of space	"It's as if the piece explores dichotomies of form/space, sound/silence." (PNU)
		"A sudden shift from small-large, narrow-wide." (UT)
	Silence as space	"The drone slowly increases in intensity until it reaches a halt." (PNU)
		"Space is achieved by silence" (TCCTF)
	Spatial practices	"It reveals the tail of the reverberation that was present all along." (PNU)
Patterns are mentally constructed	Grasping to memory and anticipation	"As [the drones] gained prominence, the main sequence went to the background, yet I was clinging to its familiarity. Upon its return, I was almost relieved." (BU)
	Harmonic experience	"The lack of 'obvious' harmonic centre for the piece takes me away from listening to music in a certain way." (MPVV)
	The role of repetition	"I was confused by the sequence, yet I kept on listening as it started to exhibit certain discernible tones. The sequence starts revealing itself." (TLL)

(2008) illustrates the exploration of these constituents. The piece incorporates the voice of Harvey's son and the sound of a bell. As the composition progresses, one can discern different practices of construction and fragmentation facilitated by digital technology. These practices reveal the fundamental building blocks of sound, like the individual 'hit' and resonating partials produced by the bell.

Bells connote the experience of a meditation retreat, as gongs are used to notify meditators about different aspects of the schedule, like a beginning or an end of a meditation. Yet, their manipulation in this piece brings forth a certain insubstantiality. This notion is explored in-depth by American Buddhist teachers, Goldstein and Kornfield:

> If a bell is rung, what do we hear? Most people hear a 'bell', or if there's a noise outside, might say that we hear a car or a truck going by. But that's not what we hear. We hear certain sounds, certain vibrations, and then immediately the mind names it as 'bell', 'car', 'truck', or 'person'. We confuse the concepts of the thinking mind with the reality of direct experience. (2001, p. 25)

The focus on the 'reality of direct experience' corresponds with a fascination with timbre. Whereas previously, timbre was associated with material transformation, here it is seen as a constituent within a constellation of auditory perception. This is reflected in the KYE entry in Table 6.2. Timbre and its associated terms, harmonics and spectra, alongside additive/subtractive process of synthesis and the example of the bell, seem to point toward the paradoxical question 'What constitutes a sound?', challenging our perception of fixed concepts.

6.7.2 Subtheme 3.2: Patterns Are Mentally Constructed

During initial analysis, a cluster of meaning was formed around the concept of 'patterns'. Initially, segments were coded semantically, but as the analysis progressed, this code also developed into a subtheme. The label '*Patterns are mentally constructed*' represents a multifaceted concept that includes the role of musical repetition in establishing patterns, the workings of memory and anticipation, and their connection to familiarity and attachment. It also ponders a nondual approach to the question 'What constitutes a pattern?', according to which our experience of patterns results from a simultaneous interaction between a listening subject and a sounding world.

Patterns are mentioned throughout the dataset, often as a sequence of pitches organised in time. Sequences form melodies, but they can be expanded to include harmonic movements. Regardless of what defines them, patterns are established in a process of repetition or 'looping' as an affordance of our memory (Margulis, 2014). Contrasting past iterations with current ones, we might note *similarity* or *difference*. Variation or melodic difference is seen in several journal excerpts, causing "familiarity with a sense of shifting" (DA) and "intrigue" (TCCTF). Whereas prolonged similarity in pitch can often facilitate a shift of attention toward other, less obvious elements—timbral or other subtle nuances.

Another aspect of our experience of repetition can be gleaned through a segment from BU. There, we see the role of repetition in the formation of unpleasant 'clinging' and the pleasant 'relief' that follows it. This certainly relates to memory, but often it is caused by anticipation of a 'crescendo' that connotes the all-too-familiar riser of Electronic Dance Music. However, the most pertinent aspect of this theme suggests the role of the perceiver in the formation of patterns. When listening to the track 'The Landscape Listens' by Barbieri (2022), the pattern, which is often conceived as the essence of the sound object, is experienced as transitioning from an undefined state to a more distinct one:

> TLL: [My mind was] trying to grasp a pattern in the pitch, yet its materiality is in constant transformation [. . .] That which constitutes the pattern is never obvious, it is empty of substance.

This quote emphasises the notion that patterns are mental constructs of tonal relations, insubstantial when held against the scrutiny of present-moment experience.

6.7.3 Subtheme 3.3: A Sense of Space

While the subthemes above suggest the insubstantiality of music by considering its constituents and patterns, this subtheme does so via the interaction between sound and space. On an apparent level, the journals repeatedly refer to an appreciation of space and an awareness of different spatial practices in the works of other composers. Those excerpts were originally organised under a standalone theme, labelled '*A sense of space*'. However, since they imply a nondual relationship between sound and the space it inhabits, they were eventually nested under the theme on emptiness. To explore the different aspects of this subtheme, it is worth introducing the idea of space as a material encounter.

In his essay 'The Materiality of Space', Nelson (2015) challenges the conventional associations of materials with solidity and spaces with emptiness. By considering a material encounter between sound and space, Nelson rejects the idea of space as an abstract and neutral geometry. Drawing from social and ecological theories, he argues that space is actively produced by human action and is present in an ontology of transformation characterising materials and their interactions. Aligning this perspective with the Buddhist doctrine of *emptiness* disrupts the notion of space as a stable object, emphasising its dynamic nature influenced by social, political, physical, and ecological factors in continuous development. The exploration of the spatial development of sound suggests a reciprocal relationship between sound and space. Sound is not merely present in space; space is actively explored by sound. When we appreciate space as imbued in and explored by sound, we appreciate sound and space as complex phenomena that exists in a nondual relationship, both separate and unified.

Nelson proposes that proximity or distance in sound mirrors the compactness or expansiveness of space, achieved through practices of amplification and attenuation. Loudness implies presence, while distance suggests separateness or longing. The juxtaposition of sound and silence becomes a spatial practice, invoking subjective notions of proximity and historicity. This relationship is exemplified in excerpts from TCCTF and PNU, where contrasts between sound and silence elicit reflections on what was once present and is now absent. Much like directing attention at itself, spatial practices can promote nondual states of background awareness, and this will be discussed as a form of transcendence in the next theme.

6.8   Theme 4: Transcendent Experiences With Music

Based on transcendence, as a feeling of departure from one state of being to another, this theme was constructed around segments that refer to profound epistemic leaps promoted by listening to musical works. Labelled '*Transcendent experiences with music*', this multifaceted concept is situated within the works of Gabrielsson (2011), Becker (2004), and Clarke (2014). However, it is shaped by my experience as a meditation practitioner and therefore influenced by Buddhist thinking. It was assigned to segments that note a mysterious/otherworldly/eternal presence, a connection to other people, and a feeling of immersion or blurring between self and world. These facets are seen in Table 6.5, alongside example excerpts from the journals.

Multiple journal segments refer to feelings of connection to something greater, mysterious, or eternal. Often inspiring awe and intrigue, these experiences seem to support Gabrielsson's (2011) findings on strong experiences of music. One journal entry alludes to a further aspect, and while it cannot be considered a pattern on its own, it is nevertheless worth mentioning. During the summer of 2022, I attended an electronic music festival in Crato, Portugal. There, a performance by artist Abul Mogard felt particularly significant, and I decided to document it in a listening journal (LAWL). While the other journal entries suggest a connection with an otherworldly entity, the excerpt from LAWL discusses a distinct type of connection—human connection. It reflects an experience reminiscent of metta bhavana, also known as lovingkindness meditation, where practitioners strive to cultivate compassion towards themselves and others. My experience of metta bhavana has been transformative and therapeutic, evoking a sense of softening and upliftment. The power of metta bhavana meditation is amplified when practiced collectively, and the scarcity of similar experiences in the journals may be attributed to the fact that all other entries were written in solitude rather than public settings.

The third aspect of this theme refers to a feeling of transcending one's physical boundaries and sense of self. This form of transcendence also manifests as sound causes strong physical sensations or contributes to a feeling of 'immersion'. Immersion should not be conflated with immersive-audio formats, which feature multiple speaker configurations or binaural audio. Here, immersion is associated with a lack of dualistic perception, namely a collapse of subject-object experience, as seen in segments from SORTS, LAWL and KYE. During immersive experiences, we do not feel like a subject *looking at* experience. Rather, we experience *oneness* or *submerging* with the totality

TABLE 6.5 Coded segments for theme 4

Code	Example Journal Extracts
A connection to something mysterious/ divine/ other-worldly/ eternal	"I feel a connection to something bigger, more eternal, that was there before me and will exist after me" (KYE)
	"This causes intrigue [. . .] I feel as I am participating in the construction of this mystery." (TCCTF)
	Enigmatic, oscillating sound. Immediate yet eternal. (MPVV)
	The piece evokes holiness in me. A feeling of divine presence. (PNU)
	"This universal mystery humbles me. We think we know the causes and conditions of ourselves, the world, but we truly don't." (INTCAEB)
	"[this] adds to the mystique" (VW01)
A connection to other beings	"My heart opened up. I felt elevated and sensed all humans around me." (LAWL)
	Opening my eyes, I looked around and noticed the looks on people's faces. I realised they went through a similar, transformative experience. (LAWL)
Blurring the boundaries between self and experience	"It is immersive and can be said to take me away from myself, much like the sound of 'ohm'." (PNU)
	"In those moments, I felt there was no difference between me (subject) and the sound (object) [. . .] There was a sense of oneness with sound" (LAWL)
	"In moments of deep concentration, I felt immersed in sound." (MUT)
	"There were moments of pure awareness to sound, where there was no difference between the sound and myself [*sic*]." (SOTRS)
	"Like watching the waves of the sea slowly approaching, and letting it immerse you and feel how it slowly goes away." (SOTRS)
	"The piece starts with textured rain which makes me feel calm and safe, but also totally immersed." (VW01)
	"In several moments the external and internal are indistinguishable. I feel immersed." (KYE)

of experience. This nondual aspect of listening can manifest when we direct attention at itself, or when we contemplate the paradoxical relationship between sound, space, and silence. A Buddhist interpretation of this experience enables us to view it not merely as an alteration of an enduring subjectivity, but rather as a momentary glimpse into the truth of non-self, or 'mere awareness' as the Buddha calls it.

6.9 Conclusion

Examining whether listening to modular-based works can provide a window into contemplative insights associated with meditation, this study shows that deep listening can provide an alternative path for generating data, and should be considered a method for self-study. The themes generated as part of this study enter a discourse concerning listening as a focus of attention, demonstrating that a phenomenological difference is gleaned when we immerse ourselves in sound. They also tell a story of how musical works can communicate ideas like impermanence and emptiness through an emphasis on process, repetition, and a whole host of material and spatial practices.

As listeners focus their attention on the stream of mental occurrences, they come to terms with change that transpires in musical material and their own subjectivity. In the flux of auditory experience, a contemplation on sound—reduced to its meaningful constituents—helps realise that, like all other phenomena, sound is devoid of a permanent essence. In examining instances like patterns and space, the journals explore the nondual relationship between observer and observed, the object and its background. In addition, the journals recount transcendent experiences characterised by a sense of immersion in sound. This sense of absorption or subject-object collapse is discussed by other authors, yet here it is interpreted through the Buddhist teaching of non-self and nonduality.

This investigation, focused on listening and subjective experience, lacks a description of modules and patching techniques. In addition, the themes generated are not limited to music made with modular synthesisers but extend to electronic music that utilises software tools or other, non-electronic musical forms. Where does this leave us? Examining my listening experience helps guide my own work with clearer aesthetic criteria, influencing not only how works sound like, but also the dynamics and feelings they invoke in listeners. This serves as an emotional, affective compass for working with modules and patches, driving an investigation into the way ideas like impermanence and emptiness might be expressed within a modular environment.

References

Adams, T.E., Holman Jones, S. and Ellis, C. (2021) 'Introduction: Making Sense and Taking Action: Creating a Caring Community of Autoethnographers', in Adams, T.E., Holman Jones, S. and Ellis, C. (eds) *Handbook of Autoethnography*. 2nd edn. New York: Routledge, pp. 1–19. Available at: https://doi.org/10.4324/9780429431760-1

Aldridge, D. (2003) 'Music therapy and spirituality; A transcendental understanding of suffering', *Music Therapy Today*, 4(1), pp. 1–28.

Barbieri, C. (2017) *Patterns of Consciousness*. Important Records. Available at: Spotify (Accessed: 7 August 2023).

Barbieri, C. (2018) 'Caterina Barbieri on synthesis, minimalism and creating living organisms out of sound'. Available at: www.factmag.com/2018/07/08/caterina-barbieri-signal-path/ (Accessed: 21 September 2022).

Barbieri, C. (2022) *The Landscape Listens*. Light Years. Available at: Spotify (Accessed: 7 August 2023).

Becker, J. (2004) *Deep Listeners: Music, Emotion and Trancing*. Bloomington and Indianapolis: Indiana University Press.

Biswas, A. (2011) 'The Music of What Happens', in D. Clarke and E. Clarke (eds) *Music and Consciousness: Philosophical, Psychological, and Cultural Perspectives*. Oxford; New York: Oxford University Press, pp. 95–110. Available at: www.vlebooks.com/Vleweb/Product/Index/2047239?page=0 (Accessed: 25 October 2021).

Borgo, D. (2012) 'Embodied, Situated and Distributed Musicianship', in A.R. Brown (ed.) *Sound Musicianship: Understanding the Crafts of Music*. Newcastle: Cambridge Scholars, pp. 202–212. Available at: https://go.exlibris.link/4vhMcR5R

Braun, V. and Clarke, V. (2022) *Thematic Analysis: A Practical Guide*. London: SAGE. Available at: https://go.exlibris.link/Tzp2bZRz

Bruno, C. (2016) 'An artist's approach to the modular synthesizer in experimental electronic music composition and performance', *CEC | Canadian Electroacoustic Community* [Preprint]. Available at: https://econtact.ca/17_4/bruno_modularsynth.html (Accessed: 23 February 2021).

Chang, H. (2008) *Autoethnography as Method*. Walnut Creek, CA: Oxford; Left Coast (Book, Whole). Available at: https://go.exlibris.link/Wkq6whNN

Clarke, D. and Clarke, E. (eds) (2011) 'Preface', in *Music and Consciousness: Philosophical, Psychological, and Cultural Perspectives*. Oxford; New York: Oxford University Press, pp. xvi–xxiv. Available at: www.vlebooks.com/Vleweb/Product/Index/2047239?page=0 (Accessed: 25 October 2021).

Clarke, E.F. (2014) 'Lost and found in music: Music, consciousness and subjectivity', *Musicae Scientiae* [Preprint]. Available at: https://doi.org/10.1177/1029864914533812

Cohen, L., Manion, L. and Morrison, K. (2017) *Research Methods in Education*. London: Taylor & Francis. Available at: http://ebookcentral.proquest.com/lib/uwestlon/detail.action?docID=5103697 (Accessed: 6 January 2022).

Demers, J. (2010) *Listening through the Noise: The Aesthetics of Experimental Electronic Music*. New York; Oxford: Oxford University Press.

Ellis, C., Adams, T.E. and Bochner, A.P. (2011) 'Autoethnography: An overview', *Forum: Qualitative Social Research*, 12(1). Available at: www.qualitative-research.net/index.php/fqs/article/view/1589 (Accessed: 2 March 2022).

Gabrielsson, A. (2011) *Strong Experiences with Music: Music is Much More than Just Music*. Translated by R. Bradbury. New York; Oxford: Oxford University Press.

Geismar, H., Otto, T. and Warner, C.D. (2022) *Impermanence: Exploring Continuous Change across Cultures*. London: UCL Press.

Goenka, A.S.N. (1993) *Mahasatipatthana Sutta*. Translated by A.S.N. Goenka. Igatpuri: Vipassana Research Institute.

Goenka, A.S.N. (2017) *Discourses on Satipatthana Sutta*. Third. Igatpuri: Vipassana Research Institute.

Goldstein, J. and Kornfield, J. (2001) *Seeking the Heart of Wisdom: The Path of Insight Meditation*. Boulder, CO: Shambhala Publications.

Hart, W. (2018) *Vipassana Meditation – The Art of Living*. Dhamma Giri, Igatpuri: Vipassana Research Institute.

Harvey, J. (2008) *Mortuos Plango, Vivos Voco*. SARGASSO. Available at: Spotify (Accessed: 4 August 2023).

Hetrick, M.L.S. (2017) *Modular Understanding: A Taxonomy and Toolkit for Designing Modularity in Audio Software and Hardware*. University of California. Available at: https://search.proquest.com/docview/189 2089007 (Accessed: 23 February 2021).

Hyde, J. (2016) 'It's Not an Instrument, It's an Ensemble: A Parallel approach to modular synthesizer design', *CEC | Canadian Electroacoustic Community* [Preprint]. Available at: https://econtact.ca/17_4/hyde_e nsemble.html (Accessed: 23 February 2021).

Lowe, B. (2011) '"In the Heard, Only the Heard . . . ": Music, Consciousness and Buddhism', in D. Clarke and E. Clarke (eds) *Music and Consciousness: Philosophical, Psychological, and Cultural Perspectives*. Oxford, New York: Oxford University Press, pp. 111–136. Available at: www.vlebooks.com/Vleweb/Prod uct/Index/2047239?page=0 (Accessed: 25 October 2021).

Margulis, E.H. (2014) *On Repeat: How Music Plays the Mind*. New York; Oxford: Oxford University Press.

Montague, E. (2011) 'Phenomenology and the "Hard Problem" of Consciousness and Music', in D. Clarke and E. Clarke (eds) *Music and Consciousness: Philosophical, Psychological, and Cultural Perspectives*. New York; Oxford: Oxford University Press, pp. 29–46. Available at: www.vlebooks.com/Vleweb/Prod uct/Index/2047239?page=0 (Accessed: 25 October 2021).

Navs (2016) 'Basic Electricity: An appeal for a greater understanding of rudimentary modular functions', *CEC | Canadian Electroacoustic Community* [Preprint]. Available at: https://econtact.ca/17_4/navs_basic electricity.html (Accessed: 6 December 2021).

Nelson, P. (2015) 'The materiality of space*', *Organised Sound*, 20(3), pp. 323–330. Available at: https://doi. org/10.1017/S1355771815000254.

O'Connor, N. (2019) 'Reconnections: Electroacoustic Music & Modular Synthesis Revival', in. Electroacoustic Music Association of Great Britain at University, University of Greenwich, UK . . .

Oliveros, P. (2005) *Deep Listening: A Composer's Sound Practice*. New York: iUniverse.

Priest, G. (2009) 'The structure of emptiness', *Philosophy of East and West*, 59(4), pp. 467–480.

Radigue, E. (1994) *Trilogie De La Mort*. XI Records. Available at: Spotify (Accessed: 9 January 2022).

Radigue, E. (2009) 'The mysterious power of the infinitesimal', *Leonardo Music Journal*, 19, pp. 47–49.

Reich, S. (2017) 'Music as a Gradual Process', in C. Christoph Cox (ed.) *Audio Culture: Readings in Modern Music*. 2nd edn. London: Bloomsbury Academic.

Styles, E. (2006) *The Psychology of Attention*. 2nd edition. New York; Hove: Psychology Press. Available at: https://ereader.perlego.com/1/book/1604404/11?element_originalid=ncx_84 (Accessed: 19 December 2022).

Varela, F.J., Thompson, E. and Rosch, E. (2016) *The Embodied Mind, Revised Edition: Cognitive Science and Human Experience*. Cambridge, Massachusetts; London: MIT Press.

Wahbeh, H. *et al.* (2018) 'A systematic review of transcendent states across meditation and contemplative traditions', *EXPLORE*, 14(1), pp. 19–35. Available at: https://doi.org/10.1016/j.explore.2017.07.007.

Wu, J.C. (2020) 'From physical to spiritual: Defining the practice of embodied sonic meditation', *Organised Sound*, 25(3), pp. 307–320. Available at: https://doi.org/10.1017/S1355771820000266

Discography

Barbieri, C. (2017) *Patterns of Consciousness*. Important Records. Available at: Spotify (Accessed: 7 August 2023).

Barbieri, C. (2022) *The Landscape Listens*. Light Years. Available at: Spotify (Accessed: 7 August 2023).

Harvey, J. (2008) *Mortuos Plango, Vivos Voco*. SARGASSO. Available at: Spotify (Accessed: 4 August 2023).

Radigue, E. (1994) *Trilogie De La Mort*. XI Records. Available at: Spotify (Accessed: 9 January 2022).

7

RESEARCH THE EFFECT OF VISUAL STIMULI ON AUDITORY PERCEPTION IN MUSIC RECORDING AND LISTENING

Pengcen Liu

7.1 Introduction

It has been well established that visual perception is dominant among the senses (Ryan, 1940). There are already many documents which demonstrate the influence of vision on auditory perception [e.g., Kato and Kashino (2001), Saldaña and Rosenblum (1993)]. This is an established fact that the auditory system can be affected by visual perception (Howard and Angus, 2009, 'Acoustics and psychoacoustics'). With the advent of the use of computer software to make audio recordings, the dependence on visual information to make changes to the sounds may result in tricking the ear into thinking changes have occurred, when in fact, they have not. When information from several sensory modalities clashes with visual information, vision often "wins" (Guttman et al., 2005).

For example, I witnessed a sound engineer made changes using a visual interface on Channel 1 and he thought that there was a difference in the sound. But he did not realise he was changing the sound on Channel 2. This experience made me want to research, in more depth, the dependency on visual information in the manipulation of audio. At the same time, seeing a visual cue can assist us in focusing on a particular detail within the sound.

Nowadays, many scholars are studying vision-auditory systems more deeply than before. And most of the research has been done by psychologists and linguists. The "McGurk effect" (McGurk and MacDonald, 1976) provides strong evidence for the influence of sight on auditory perception. Some scholars have developed it and tried to expand this effect from the field of speech to the field of music. These studies also promoted the development of visual-auditory research in the field of music.

McGurk and MacDonald published 'Hearing lips and seeing voices' (1976). Mentioned in the research is that when people hear a syllable like 'bah' but people see the shape of a mouth is a syllable like 'gah', then people will hear the third syllable 'dah'. The effect is especially strong when the auditory signal is degraded (Sekiyama and Tohkura, 1991). This phenomenon is known as the McGurk effect. Besides the McGurk effect, the bouba-kiki effect (Nathan and Laurent, 2019) can also prove the connection between vision and hearing. However, the McGurk effect remained in the realm of speech, not involving the field of music. Many scholars explored this discussion after McGurk and MacDonald. 'Visual influences on auditory pluck and bow judgments' was written by Saldaña and Rosenblum (1993). They tried to verify the influence of vision on auditory in testing

DOI: 10.4324/9781003396710-7

speech, and then further research into musical instruments. In their first experiment, they recorded a plucked (pizzicato) then a bowed (arco) cello, and then the dubbed audio files were synced with the video displays. Thirteen students made each identification judgment. Saldaña and Rosenblum were effective in showing that visual cues have an impact on auditory identification decisions made in response to natural non-speech stimuli. But compared to speech effects, this non-speaking impact is far less noticeable (Experiment 2).

Although it can be proved that auditory perception is influenced by vision in the McGurk effect, the auditory cortex's ability to interpret phonetic information may be impacted by early, extensive musical instruction. 'Skilled musicians are not subject to the McGurk effect' (Proverbio et al.,2016) However, Politzer-Ahles and Pan found the limitations of Proverbio and colleagues' finding. They discovered that musicians do exhibit a significant McGurk effect that is both larger than that of non-musicians and statistically significant. 'Skilled musicians are indeed subject to the McGurk effect' (Politzer-Ahles and Pan, 2019). The results of the latter suggest that further research focused on musicians is feasible. In particular, research that explores how visual information can improve musicians' hearing should be explored. Spatial audio-visual links also exist (Kato and Kashino, 2001) The results of their current study show that the introduction of the visual-spatial cue greatly increases not only the decision criterion but also the sensitivity of auditory spatial discrimination. This conclusion also demonstrates that the visual can improve the sensitivity of the auditory in some respects. What makes this argument even more compelling is that the auditory threshold of the human hearing in the masking effect will decrease in response to visual cues (Chen and Xie, 2010). Chen and Xie demonstrated another way in which visual supports auditory. The argument and experiment of Chen and Xie are similar to this research project. However, it does not involve music production, and it does not involve music technology.

This research is an exploration of visual-auditory perception in music post-production, which is based on the research results of visual-auditory in physiology, psychology and linguistics. Therefore, it is inevitable that other aspects of the subject will be mentioned.

7.2 Method

7.2.1 Participants

Eighteen participants took part in the experiment. Five participants major in piano, three participants are from music technology, one participant is a violinist, one is a jazz drummer and the other eight participants are not learning any music. Ten participants are male, and eight participants are female. Their average age ranges from 20 to 25. All participants have a normal hearing system and judgment ability. Ethical approval for the experiment was obtained from Birmingham City University.

Participants include non-musical, musicians and students from music technology who are thought to be more familiar with recording studios and music post-production. The consequence of the experiment can be more objective, compared with using monotypic participants. However, the study is conducted for music post-production, so it is debatable whether all the participants should be music engineers. There are few studies about untrained listeners' and trained listeners'(musicians) perceptions of pitch and timbre (Pitt, 1994). However, there has been no paper comparing perceiving variation of pink noise frequency between ordinary people, musicians and music engineers.

7.2.2 Apparatus

The experiment took place in the Mastering suite at the Royal Birmingham Conservatoire. This room consists of 2 Bowers & Wilkins 800D3 speakers, which were used as experiment speakers. Output sound level was controlled by a Maselec MTC-1X. All videos were displayed on a Samsung computer screen that was connected to a MacBook Pro with a HDMI cable. All the audios were made by Pro Tools which generated pink noise and took responsibility for audio automations. The variation of frequencies was controlled by Fabfilter Pro Q3, even though some videos contain the interface of Waves SSL EV2 Channel.

7.2.3 Preparation

Before the formal experiment, participants were asked to listen to different frequency parts of the pink noise, 64 Hz, 125 Hz, 250 Hz, 500 Hz, 1k Hz, 2k Hz, 4k Hz, 8k Hz and 16k Hz, till their hearing systems were able to recognise the loudness of these frequencies increasing. The reason why they need to recognise the different frequencies is that it was found in a preliminary experiment that people who had never heard pink noise had difficulty hearing the loudness of a particular frequency increasing in the noise when they were first exposed to it. All of 18 participants can not perceive the 64 Hz increasing at the first time they listen to pink noise in the studio, but they can hear the level of 64 Hz increasing clearly after they listen to different frequency parts. During this training, participants were asked to listen to 125 Hz, 1k Hz and 8k Hz three times to acquaint participants with these special frequencies that would be used in the formal experiment.

7.2.4 Experiment Instruction

There are two experiments in this paper. Each one contains visual stimuli and auditory stimuli. The first experiment aims to research whether visual stimuli will reduce the threshold of participants hearing the frequency changing (the increased loudness of a specific frequency). Several representative frequency segments from low frequency to high frequency are selected: 60 Hz, 250 Hz, 1k Hz, 4k Hz and 12k Hz. Each video is 16 seconds long. The change that is modified by Fabfilter Pro Q3 appears at the third second. The parameter is promoted from 0 dB to 12 dB in 10 seconds, which means the particular frequency will be promoted to the highest (12 dB) at the thirteenth second, and then continue for 3 seconds (Figure 7.1).

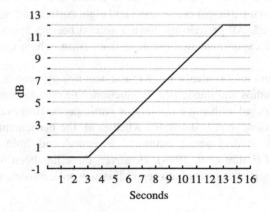

FIGURE 7.1   Volume changing trend curve.

The EQ bandwidth (Q) (the specific value is not found in the manual) is default and coincident during the whole experiment. The interpretation of the Q values might not be the same with other EQ plugins (Fabfilter Pro-Q3 User Manual). Because Q does not change crossing all experimental fragments, the interpretation of Q is not important to this experiment. Each different frequency segment was tested in two forms, one with a visual stimulus form (video) and one without video. Participants were asked to stop the playing by pressing a space on the keyboard once they heard a change in the specified frequency, whether there was a visual cue or not.

The second experiment contained two research objectives. The first goal is to investigate if auditory judgement will be inaccurate when exposed to visual stimulus with EQ plugin, or an illusion that is false. The second goal is to investigate if people's auditory judgement is influenced differently by the visual inputs of various display interfaces (different EQ plugins). Choose from two of the mixing industry's most illustrative EQ effects (Waves SSL EV2 Channel and Fabfilter Pro Q3).

There are two main categories of EQ effectors plugins which are popular (Shaurya Bhatia (2022), Top 14 EQ Plugins for Mixing 2023) nowadays. The first type of EQ effectors plugin is a recreation of the legendary SSL 4000E console channel strip (Waves website). Their characteristic is that there is no spectrum on the board of effector with only a few fixed frequency bands. The other type of EQ effectors plugin is more like a modern digital effector with intelligent interface (Fabfilter Pro-Q3 User Manual). Their primary characteristics 20 Hz to 20k Hz frequency responsiveness can be represented as a curve. On this curve, any frequency bands may be picked and altered, and there is a real-time spectrum that varies with the audio. This experiment selected two representative effector plugins, which can accurately represent two distinct visual cues.

Three frequency segments which can represent low (20–200 Hz), middle (300 Hz-5k Hz) (Svsound.com) and high frequency (8k Hz-16k Hz are selected: 125 Hz, 1k Hz and 8k Hz. All tests were based on these three representative frequencies. The experiment's visual cues come in three different variations: none (Figure 7.2). Despite this, participants still need to glance

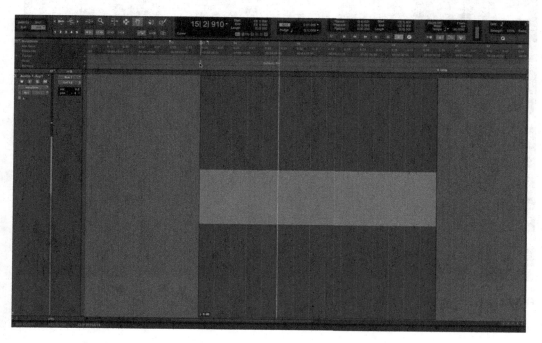

FIGURE 7.2   Visual cue: None.

FIGURE 7.3 Visual cue: Waves SSL EV2.

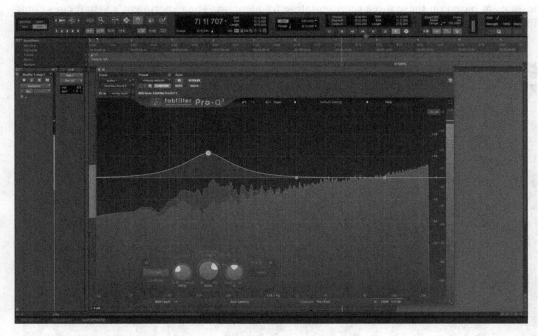

FIGURE 7.4 Visual cue: Fabfilter Pro Q3.

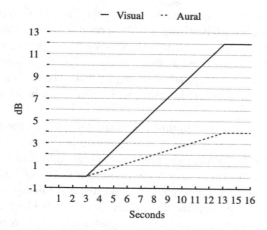

FIGURE 7.5   Visual value and real value.

at the computer. There is a Pro Tools audio clip in this video; the Waves SSL EV2 effector plugin interface cue (Figure 7.3), often known as the analogue console interface visual cue; and the Fabfilter Pro Q 3 effect interface cue (Figure 7.4), also known as the intelligent spectrum prompt.

These three forms of visual stimuli are increasingly visualised, ranging from no visual cue to adjusting a single knob on a digital analog mixer plugin interface to raising a frequency segment in the spectrum. There are two kinds of EQ plugins in the experiment, and their effects are different in terms of timbre, loudness enhancement, and Q value. If the audio is completely processed according to the plugin shown in the video, then the experiment will be inaccurate. Because there is difference between increasing 4 dB at 1k Hz on Waves SSL EV2 by its default Q value and increasing 4 dB at 1k Hz on Fabfilter Pro Q3 by its default Q value. Considering that the effect of each plugin in this experiment is not important, what is important is the comparison of experimental data under multiple groups in the same situation, so the EQ effect of Fabfilter Pro Q3 is used as the actual tool to change the experimental audio in all experiments. Even while in the video the participant sees the knob on Waves SSL EV2 Channel being improved, the Fabfilter Pro Q3 is the one who really makes the adjustments to the audio. This eliminates the need to use plugins whose effects are distinct from one another.

To accentuate the effects of visual stimulation, the volume rise in the visual stimulation videos is significantly more than the real volume increase, which is 12 dB. In some mixing processes, engineers adjust the parameters on plugins with subtle effects to the maximum value (8 or 10 dB) and then drop them back gradually (Owsinski, 2017). Engineers will inevitably make auditory judgments under the visual stimulation of the maximum value parameters, therefore raising the visual stimulation by 12 dB can effectively imitate this scenario. The real value of the aural loudness boost is 4 dB (Figure 7.5).

The reason why 4 dB is chosen as the boost amount is because the result data of Experiment 1 shows that the average hearing threshold of the participants for each frequency is around 4 dB. Using 4 dB as the auditory parameter can simulate the engineer's processing of the auditory parameter that does not change significantly in music post-production.

Each video is 16 seconds long. The visual change that is modified by Fabfilter Pro Q3 appears at the third second. The parameter is promoted from 0 dB to 12 dB in 10 seconds, which means the frequency will be promoted to the highest (12 dB) at the fourteenth second, and then continue for 3 seconds. These changes occur identically on the interface of the two different plugins.

There are two situations in the visual stimulation experiment. The first is that the loudness of frequency in the audio is increasing while the plugin knob is raising, and the second is that the audio does not change while the plugin knob is raising. There are also two situations in the experiments without visual signals, the one with and the other one without auditory changes. The consistency of experimental circumstances is ensured by the fact that there are changing frequency bands in the audio even in the absence of visual cues, and participants are informed which frequency bands in the audio they are about to hear will change possibly.

Before pink noise was used as the test audio source, three multi-track music songs were chosen as a test audio source. Three parts of each song were chosen as test points: drums, guitar and vocal and EQ adjustments (same as the formal experiment) were made to the high, mid, and low frequencies of the three parts, for example, the guitar part of song A was boosted at 1k Hz for 4 dB, the drum part of song C was boosted at 125 Hz for 4 dB, and so on, then requested that participants determine if the sound source had changed or not. The experiment's underlying logic was identical to the formal experiment, but the sound source grew more complex. After a preliminary experiment with several participants, it was determined that, due to the song's extreme intricacy, it was difficult for the participants to form judgements using additional visual and auditory clues. They cannot recall the timbres of three parts in each of the three pieces, totalling nine parts; hence, even if the change had obviously occurred or had not occurred, the participants may erroneously conclude that it was the song itself changing rather than the sound which was modified by the test effectors. In the second pre-experiment, therefore, the samples of the experiment were decreased from three frequency band tests of nine voices of three songs to three frequency band tests of three voices of one song. Although the complexity of the experiment had been reduced, the participants were still confused, and they still could not judge whether the EQ changes they heard were changes in the song itself or by effectors on the screen in the experiment. The reason is that none of the participants had heard the test song, and even if they had, they could not remember the details of a certain segment of the song. Therefore, using songs as the experimental sound source failed, because it contained uncontrollable variables in the experiment, which affects the accuracy of the experiment. The selection of pink noise may accommodate the high, medium, and low frequencies of human hearing and ensure that the participants' subjective perception of the sound source remains unchanged.

In two experiments, participants were instructed to look at the screen displaying the visuals whenever any audio was playing, even if there was nothing playing on the screen, listening with eyes closed for long periods of time was forbidden, with the exception of blinking, which is designed to prevent more uncontrollable factors appearing, despite it been proven that closing eyes will not enhance auditory attention, but fewer variables will make the experiment simpler and more accurate (Wostmann et. al, 2019). All participants sat in the same position. The size of the visual cue window was the same. The sound level was controlled by a Maselec MTC-1X as -32, which means participants received the same sound pressure level as each other.

Before playing the visual stimulus videos, participants were told what frequency segment they were about to hear. The purpose of this is to allow participants to focus on the frequency bands that need attention in advance. To prevent the participants' attention from being placed on other frequency bands that have not changed, if this happens, even if the target frequency band has changed, the test results may be inaccurate due to divided attention. In a divided-attention task, you try to pay attention to two or more simultaneous messages, responding to each as needed. In most cases, your accuracy decreases, especially if the tasks are challenging (Ward, 2004).

Testing videos appeared randomly, which could prevent the participants from perceiving the regularity and reduce the experimental error.

7.3 Result

7.3.1 Experiment 1

The main information can be obtained from the following two charts (Table 7.1, Graph 3). There are some details that are not shown in the chart and are presented separately.

Two people exhibited lower auditory thresholds without visual cues than with visual cues during the 60 Hz auditory threshold test, while the results of the remaining 16 participants indicated that visual cues at 60 Hz decreased hearing threshold. In the 250 Hz auditory test, one of the eighteen participants had a lower auditory threshold without visual cues than with visual cues. However, the findings of the remaining seventeen people indicated that visual cues decreased the 250 Hz auditory threshold. The data condition at 12k Hz and 250 Hz is the same. At 4k Hz, one participant's hearing threshold did not diminish in the presence of visual stimuli, but another participant's auditory threshold remained unchanged in the absence of visual signals.

Each frequency band's data is averaged and transformed into a graph. It can be observed from the graph (Figure 7.6) that each curve is in line with the Equal-loudness contour (Graph 6), indicating that there is no outcome in the experiment that contradicts the accepted facts. Comparing the values of the two curves reveals that the human ear's hearing threshold has dropped in response to the visual stimulation of Fabfilter Pro Q3. Under such visual cues, the individuals were more likely to hear the frequency variations.

7.3.2 Experiment 2

The Experiment 2 data is shown in the chart (Table 7.2). (The data in the graph (Figure 7.7) was obtained by taking the average of the correct rate of three frequency bands).

TABLE 7.1 Result data with/without visual cue

Frequency (Hz)	With visual cue (dB)	Without visual cue (dB)
60	4.366	5.974
250	3.911	5.291
1000	3.773	4.747
4000	3.464	4.306
12000	3.904	4.935

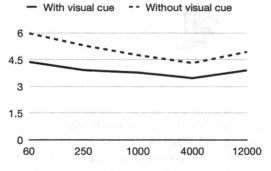

FIGURE 7.6   Result data in line graph.

TABLE 7.2 Experiment 2 result

Visual Type	Frequency	Audio Change	Rate
Fabfilter Pro Q3	125	No	11.11%
Waves SSL	125	No	50%
No visual cue	125	Yes	50%
Waves SSL	125	Yes	77.78%
Fabfilter Pro Q3	125	Yes	100%
Fabfilter Pro Q3	1000	No	22.22%
Waves SSL	1000	No	55.56%
No visual cue	1000	Yes	66.67%
Waves SSL	1000	Yes	72.22%
Fabfilter Pro Q3	1000	Yes	100%
Fabfilter Pro Q3	8000	No	27.78%
Waves SSL	8000	No	55.56%
No visual cue	8000	Yes	66.67%
Waves SSL	8000	Yes	77.78%
Fabfilter Pro Q3	8000	Yes	88.89%

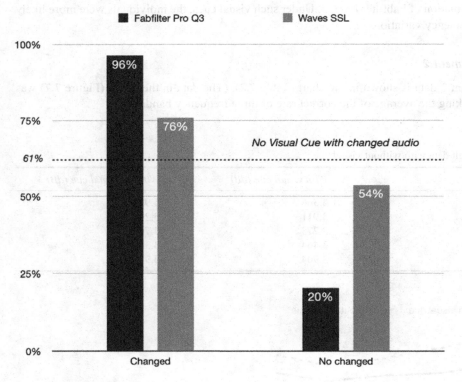

FIGURE 7.7 Accuracy rate.

Participants' aural accuracy was greatest across all three frequency bands when the Fabfilter Pro Q3 was used as a visual cue and the audio was altered. When Waves SSL was used as a visual cue and the audio was altered, the auditory accuracy of the participants was lower than when Fabfilter Pro Q3 was used as a visual cue, but higher than when there was no visual cue. This again demonstrates the conclusion of Experiment 1's findings, namely that visual cues make it simpler for participants to detect auditory changes. It also demonstrates that Fabfilter Pro Q3 as a visual signal makes people more auditory than Waves SSL.

When Fabfilter Pro Q3 was used as a visual cue and the audio did not change, the participants' auditory accuracy was the lowest of the three frequency bands; that is, with this visual cue, the participants believed the audio had changed even when it had not. The participants' accurate rate was much higher when Waves SSL was used as a visual cue than when Fabfilter Pro Q3 was used as a visual cue, but marginally lower when there was no visual cue auditory change. When participants listened to 125 Hz without a visual cue but with audio changes and without audio changes but with Waves SSL, they were equally accurate.

One participant out of eighteen accurately judged the unchanged audio under the visual stimulation of Fabfilter Pro Q3 in the three frequency bands while listening to the unchanged audio. This individual has no music background. In the auditory test of three frequency bands, three people listened to unchanged audio under the visual stimulus of Waves SSL, and they all made accurate assessments. Similarly, these three participants have no music background. One of them, the same individual stated previously, made accurate judgements on all six items of the visual distraction test. Consequently, Fabfilter Pro Q3 visual stimuli caused more auditory distraction than Waves SSL did.

Data that is not shown directly in the chart: 10 participants with a music background (including students of music technology) have a 63% overall accurate rate; 8 participants without a music experience have a 65% overall right rate; and 3 students of music technology have a 55% overall correct rate. Ten participants with a background in music listened to unchanged audio under two different visual cues. Ten of them made a total of 21 right judgements, of which 6 were made in response to the Fabfilter Pro Q3 visual stimulus and 15 were made in response to three Waves SSL visual stimuli; the overall rate of correctness is 35%. Eight participants without a musical background listened to the audio without change under two visual cues. Eight of them made 19 valid judgements in total. Among them, 5 times were accurately rated under the visual stimulation of Fabfilter Pro Q3, and 14 times were accurately judged under the visual stimulation of Waves SSL, and the overall accuracy rate was 40%.

7.4 Discussion

The first experiment's result graph demonstrates that using Fabfilter Pro Q3 as a visual stimulus can decrease the human hearing threshold of frequency perception. However, the graph reveals that the influence of visual cues is comparatively highest at 60 Hz and relatively least at 4k Hz. A visual cue efficiency curve is created by subtracting the initial auditory threshold from the auditory threshold under the visual cue. (Figure 7.8) The basic tendency of the visual cue efficiency curve is comparable to that of the Equal-loudness contour (Figure 7.9) (Fletcher and Munson, 1933), as shown by the curve.

This may indicate that visual cues are less efficient in the frequency range where the human ear is more sensitive. Thus, as people's conviction in their hearing drops, the need for visual assistance will rise.

A further verification of the result reached in Experiment 1 was conducted in the second experiment, which also investigated the impact of a variety of visual stimuli (various EQ effectors) on this basis. The findings of the experiment demonstrated that the auxiliary effects of visual signals on hearing are strengthened to the degree that there is a greater quantity of visual stimulation. In addition to serving as help for the participants' hearing, the figurative visual stimuli also misled

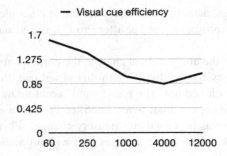

FIGURE 7.8 Visual cue efficiency curve.

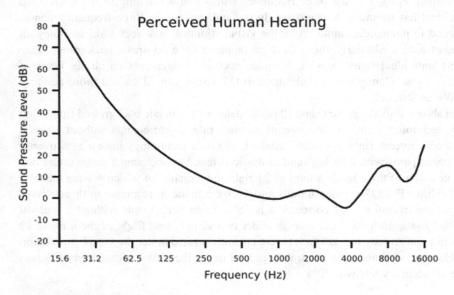

FIGURE 7.9 Equal-loudness contour.

most of the participants, since they reported hearing changes in the audio that were not actually there. The results of this study suggest, although in a roundabout way, that when music engineers modify the EQ effectors, not only will they be impacted by aural stimuli, but visual signals will also be involved in the judgements and decisions that they make.

The participants in this experiment who either have or do not have a background in music did not exhibit a significant difference. According to the collected statistical information, the percentage of correct judgements made by participants who do not have a background in music is slightly higher than that of participants with a musical background, while the percentage of correct judgements made by participants with a music technology background is lower than the average. (The sample size of participants is too small to meet the conditions for a separate classification discussion. As a result, this paper will not discuss the data results of participants with a background in music technology; rather, they are placed in the same category as participants who have a musical background.) This paper is unable to draw conclusions regarding whether individuals with a musical background are more likely to be affected by visual stimuli. Because this paper does not set specific research conditions for the study of different groups of people, the reason why two

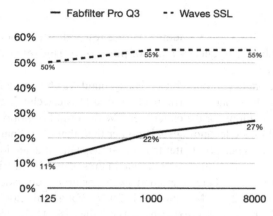

FIGURE 7.10 Statistics published by Radio Joint Audience Research (RAJAR, 2023) released in February 2023 show that 5 million people listen to Classic FM every week.

types of people (those who have no music background and those with a music background) were invited to participate in the experiment is to reduce the error caused by the different groups of people in the experiment (to research the influence of vision and hearing).

The correct rate for the three frequency bands in the presence of visual stimuli but no change in audio is graphed (Experiment 2). Even though there are only three frequency bands samples tested, the graph is also related to the visual stimulus efficiency curve (Figure 7.8). The Equal-loudness contour diagram (Figure 7.9) shows that 125 Hz is the frequency range where the human ear is the least sensitive out of the three. From the visual cue efficiency curve, we can see that this frequency range is where visual stimulation works best, while the effect of visual stimulation is the most misleading at 125 Hz because the correct rate is the lowest of the three frequency bands (Figure 7.10).

This is also the case for 1000 Hz and 8000 Hz. So, when the human ear is in the less sensitive frequency band, the visual stimulus will have a stronger auxiliary effect and a stronger misleading effect. This means that it makes the auditory effect of the human ear more biased towards the result of visual stimulus presentation.

7.4.1 Experiment Errors Discussion

The two errors in the experiment mentioned above can be classified into one category, that is, the participants judged that the audio changed when neither the audio nor the visual stimulus changed. Since none of the participants had previously engaged in a similar hearing test, participants had different expectations for the experiment and felt varied levels of stress throughout the experiment. The psychological stress of the experiment's participants and their anticipation of the experiment's auditory results will influence the participants' auditory perception (Hoskin et al., 2014).

If Experiment 1 needs to be reproduced, this error can be prevented to a certain extent by adding a countdown to the visual prompt. For instance, displaying a three-second countdown to the beginning of the audio change on the screen in front of the participant to inform the participant that the audio is going to change. Although psychological tension and expectation cannot be controlled, this strategy may minimise mistakes within a certain range. However, the error induced by psychological factors in Experiment 2 cannot be avoided in this method, which would adversely impair the experimental results.

7.5 Further Research

Sample size and experimental conditions limit this study's findings. Future research needs more participants, and the sound source should not be the frequency range of pink noise. This research should use music fragments as the sound source.

In the experiment, the correct rate of participants without a music background was higher than that of participants with a music background than that of participants with a music technology background, which exceeded the author's expectations for the experiment. With normal cognition, the sound perception ability of musicians and music technologists is higher than that of ordinary people without any music background. "Musicians perform better than non-musicians in detecting small frequency changes with smaller error rates and a faster reaction time in music, non-linguistic tones, meaningless sentences, native and unfamiliar languages, and even spectrally degraded stimuli such as vocoded stimuli" (Tervaniemi et al., 2005; Marques et al., 2007; Wong et al., 2007; Deguchi et al., 2012; Fuller et al., 2014). But the result of this experiment is not the case. Therefore, what other factors in the experiment affected the impact of visual stimulation on hearing still deserves more in-depth research.

7.6 Conclusion and Outlook

This study found that, for the 18 individuals involved, visual stimuli influenced auditory perception in the music post-production environment, and that this influence changes with Equal-loudness contour. The actual music post-production will be affected by these visual influences. For instance, when a music engineer is adjusting a parameter with a very slight amount of change, the visual indication will assist the engineer in perceiving the consequence of his adjustment. However, if engineers rely on visual cues to detect changes when modifying auditory parameters, such changes may be invisible to listeners or extremely modest. Another situation is when the engineer, relying on visual assistance to set the mix parameters, perceives that the audio has changed, but it does not due to external factors (wrong channel, FX bypass etc.). In this situation, if there is a change, the song listener will not detect it. Visual cues can either aid or mislead the operator during recording, mixing and music post-production, making it difficult to determine if they are beneficial or detrimental. The significance of the changes made by engineers who rely on visual signals for non-visual listeners is a topic that requires further investigation.

References

Bhatia, S. (2022). Top 14 EQ Plugins For Mixing 2023 (website), available online at https://integraudio.com/14-best-eq-mixing-plugin/

Chen, Y. and Xie, L. Y. (2010). Effects of visual cue on auditory masking effect, *Technical Acoustics*, Vol.29, No.4. 10.3969~.isn100—3630.2010.04.01.

Cutting, J. E. and Rosner, B. S. (1974). Categories and boundaries in speech and music, *Perception & Psychophysics*, Vol.16, No.3, pp. 564–570.

Deguchi C., Boureux M., Sarlo M., Besson M., Grassi M., Schön D., et al. (2012). Sentence pitch change detection in the native and unfamiliar language in musicians and non-musicians: behavioral, electrophysiological and psychoacoustic study. *Brain Res.* Vol.1455, pp. 75–89. 10.1016/j.brainres.2012.03.034.

Fletcher, H., & Munson, W. A. (1933). Loudness, its definition, measurement and calculation. *Journal of the Acoustical Society of America*, Vol.5, pp. 82–108.

Fuller C. D., Galvin J. J., III, Maat B., Free R. H., Başkent D. (2014). The musician effect: does it persist under degraded pitch conditions of cochlear implant simulations? *Front. Neurosci.* Vol.8, p. 179. 10.3389/fnins.2014.00179.

Guttman, S. E., Gilroy, L. A., and Blake, R. (2005). Hearing what the eyes see: Auditory encoding of visual temporal sequences, *Psychological Science*, Vol.16, No.3, pp. 228–235.

Hoskin, R., Hunter, M. D., and Woodruff, P. W. (2014). The effect of psychological stress and expectation on auditory perception: A signal detection analysis, *British Journal of Psychology*, Vol.105, No.4, pp. 524–546.

Howard, D. M., & Angus, J. A. S. (2009). Acoustics and Psychoacoustics, 4th Edition, Oxford, UK: Elsevier Ltd.

Kato, M. and Kashino, M. (2001). Audio—visual link in auditory spatial discrimination, *Acoustical Science and Technology*, Vol.22, No.5, pp. 380–382.

Leventhall, G. (2009). Low frequency noise. What we know, what we do not know, and what we would like to know, *Journal of Low Frequency Noise, Vibration and Active Control*, Vol.28, No.2, pp. 79–104.

Marques C., Moreno S., Castro S. L., Besson M. (2007). Musicians detect pitch violation in a foreign language better than nonmusicians: behavioral and electrophysiological evidence. *J. Cogn. Neurosci.* Vol.19, pp. 1453–1463. 10.1162/jocn.2007.19.9.1453.

McGurk, H. and MacDonald, J. (1976). Hearing lips and seeing voices, *Nature*, Vol.264, No.5588, pp. 746–748.

O'Callaghan, C. (2008). Object perception: Vision and audition, *Philosophy Compass*, Vol.3, No.4, pp. 803–829.

Owsinski, B. (2017). *The Mixing Engineer's Handbook*, 4th edition, Burbank, CA, Bobby Owsinski Media Group.

Palmer, C. and Krumhansl, C. (1990). Mental representations for musical meter, *Journal of Experimental Psychology: Human Perception and Performance*, Vol.16, No.4, pp. 728–741.

Peiffer-Smadja, N. and Cohen, L. (2019). The cerebral bases of the bouba-kiki effect, *Neuroimage*, Vol.186, pp. 679–689.

Pitt, M. A. (1994). Perception of pitch and timbre by musically trained and untrained listeners, *Journal of experimental psychology: Human perception and performance*, Vol.20, No.5, pp. 976–986.

Politzer-Ahles, S. and Pan, L. (2019). Skilled musicians are indeed subject to the McGurk effect, *Royal Society Open Science*, Vol.6, No.4, p. 181868.

Proverbio, A. M., Massetti, G., Rizzi, E., & Zani, A. (2016). Skilled musicians are not subject to the McGurk effect. *Scientific Reports*, Vol.6, p. 30423. https://doi.org/10.1038/srep30423.

Ryan, T. A. (1940). Interrelations of sensory systems in perception, *Psychological Bulletin*, Vol.37, No.9, pp. 659–698.

Saldaña, H. M. and Rosenblum, L. D. (1993). Visual influences on auditory pluck and bow judgments, *Perception & Psychophysics*, Vol.54, No.3, pp. 406–416.

Sekiyama, K., & Tohkura, Y. (1991). McGurk effect in non-English listeners: Few visual effects for Japanese subjects hearing Japanese syllables of high auditory intelligibility. *Journal of the Acoustical Society of America*, Vol.90, No.4, pp. 1797–1805.

Tervaniemi M., Just V., Koelsch S., Widmann A., Schröger E. (2005). Pitch discrimination accuracy in musicians vs. nonmusicians: an event-related potential and behavioral study. *Exp. Brain Res.* Vol.161, pp. 1–10. 10.1007/s00221-004-2044-5.

Ward, T. B. (2004). Cognition, creativity, and entrepreneurship. *Journal of Business Venturing*, Vol.19, No.2, pp. 173–188.

Wong P. C., Skoe E., Russo N. M., Dees T., Kraus N. (2007). Musical experience shapes human brainstem encoding of linguistic pitch patterns. *Nat. Neurosci.* Vol.10, pp. 420–422. 10.1038/nn1872.

Wöstmann M, Schmitt LS, Obleser J (2019a) Does closing the eyes enhance auditory attention? Eye closure increases attentional alpha-power modulation but not listening performance. *J Cognit Neurosci*. Advance online publication. Retrieved March 26, 2019. 10.1162/jocn_a_01403.

8

THE IMPOSSIBLE BOX

Building a DIY Groovebox on a $10 Microcontroller

Andrew R. Brown

8.1 Introduction

In this chapter, we explore the possibilities and constraints of crafting music using a low-cost microcontroller to construct a DIY algorithmic groovebox. Since their introduction as self-contained electronic musical instruments in the 1990s, grooveboxes have combined sound generation and sequencing capabilities into an all-in-one music-making device. They have been described as a democratising force in music composition, allowing anyone to create high-quality music at home or on the road.

The vehicle for this investigation is the design and development of the OnBoard Quadra, so named because it has four musical parts; drums, bass, chords played as arpeggios, and a lead line. By tracing the design and development of the Quadra, we delve into the potential of leveraging affordable technology to achieve a surprisingly versatile musical instrument.

The Quadra is designed and developed as an exercise in frugality and efficiency. Its status as an "impossible box" comes from its ability to extract 8 voices of real-time synthesised parts in stereo 48K audio from a $10 microcontroller that was originally designed for simpler tasks such as being an internet of things (IoT) device. Despite the unconventional microcontroller use, the Quadra serves as a simple, yet distinctive, groovebox for DIY music production, offering a unique and somewhat "grungy" character.

To evaluate the capability of the Quadra to achieve the musical results expected of a groovebox, a series of three case studies were conducted. The evaluative method was to identify videos of live groovebox performances in a variety of musical styles and to create works inspired by each of them using the Quadra. A critical review of outcomes is described in this chapter and links to audio/videos of the exemplar and Quadra performances are provided to allow the reader to make their own assessment.

8.2 The Groovebox

Grooveboxes encompass several essential elements, including a multi-part synthesiser/sampler for generating pitched and drum sounds, a multi-track music sequencer, and a hardware interface with controls for composition and live performances. Their popularity in music production traces back to the introduction of the term with Roland's MC-303 in 1996, although some features, aside from

DOI: 10.4324/9781003396710-8

the drum-machine element, were already present in the Roland MC-202 from 1983. The MC in the 202's name is derived from the term micro-composer. Prior to the MC-202, a few stand-alone portable synthesiser/sequencer combinations existed, dating back to the early 1970s, such as the EMS Synthi AKS.

The market for these stand-alone machines has developed significantly since the introduction of the MC-303, expanding quickly during the 1990s. Notable examples include the Roland MC series, Akai MPCs, Korg's Electribe series, Teenage Engineering's OP-1, and Elektron instruments. These devices have been seen as a democratising force in music composition (Théberge, 1997) empowering individuals to create high-quality music from the comfort of their home (Taylor, 2001). Davis even suggests that "the MC-505 [groovebox is] a catalyst for female and non-binary artists to shatter double standards and pave a path toward gender equity" (Davis, 2021).

As the capability of grooveboxes has expanded, they are increasingly used as the centrepiece of computer-less electronic music systems and can also control external sound devices. The Elektron devices have especially focused on this by including dedicated MIDI tracks tailored specifically for managing connected devices.

However, this chapter embarks on a different trajectory, diverging from the prevailing trend of escalating complexity in groovebox design. Instead, it presents a project that embraces simplicity and user-friendliness by utilising an affordable microcontroller to craft a DIY algorithmic groovebox. The aim is to explore what is minimally required to offer musicians an accessible and straightforward groovebox for their creative endeavours.

8.2.1 Affordances of The Groovebox

Grooveboxes have consistently been positioned as accessible, self-sufficient music-making devices. They emerged during a time when laptop computers were still relatively costly, and desktop computers were not feasible for live touring musicians. In this context, the groovebox filled a crucial niche and quickly gained popularity. Grooveboxes are implicated in the expansion of live electronic music, and their interfaces were designed for real-time control of precomposed elements. As Mämmi highlights, one of the notable strengths of grooveboxes "is their highly valued ability to enable live improvisation" (Mämmi, 2023).

Despite the widespread availability of mobile and affordable computing devices today, the groovebox remains highly sought-after among those who prefer a hands-on and streamlined experience as an alternative to the general-purpose computer. Grooveboxes are part of an equipment trend often referred to as a DAW-less music setup. While some manufacturers continue to produce cost-effective grooveboxes, more sophisticated ones can easily cost more than a laptop computer, so the use of these becomes a very deliberate decision.

Accessibility encompasses not only the affordability of hardware but also the assumptions about users' musical and technical backgrounds. A study of MC-303 users reveals that creative autonomy in the use of grooveboxes is related to the development of usage modes where users embrace the technology's limitations, turning them into strengths rather than obstacles (Tjora, 2009). Grooveboxes are typically used for creating popular electronic dance music genres. Indeed, Roland's advertising of the MC-303 described it as a device for creating "Techno, jungle, hip-hop, acid and other dance styles" (Roland Corp, 2023). The emphasis on concise, repeating sequences and quantised rhythmic steps ties these musical styles and features together, forming a potentially self-reinforcing creative feedback loop within the context of groovebox usage.

The control surface of a groovebox is an important element of its usability and appeal. The early Roland MC series featured a key trigger layout resembling a short piano keyboard, later Akai

MPCs introduced a grid of trigger-pads as an interface element to the groovebox, emphasising the drum and percussion focus of much of the music (Akai MPC, 2023). This interface has been adopted by others over time, including Roland, the Novation Circuit, and Native Instrument's Maschine series, further solidifying its influence in the groovebox landscape.

Web-based and mobile app music production tools are another route to accessibility. Web-based groovebox implementations include the Genius Home Studio website (Genius, 2023) and apps include the Groove Rider GR-16. Additionally, some hardware groovebox models have their app versions, including the Korg Electribe Wave 2 and Akai iMPC Pro. A review of web-based grooveboxes revealed that many of them used sample-based playback, were often less featured than their hardware equivalents, and may not yet be taking full advantage of the technology platforms they are built upon (Huttunen, 2021). The web-based groovebox developed by Huttunen, called Tahti (Huttunen, 2023) stands out for its impressive capabilities. Web-based tools are often affordable, but they still necessitate substantial computing power from a personal computer to function efficiently making them more akin to other software-based virtual grooveboxes. A more direct comparison with hardware grooveboxes might be a groovebox app running on a dedicated mobile device, such as on the Apple iPad mini.

The OnBoard Quadra strives to offer a more liberating music-making experience by incorporating generative algorithms as starting points and sources of inspiration for drum patterns, bass lines, and other elements. The stylistic variety of these generative patterns ranges from steady rock beats to chaotic Autechre-like patterns. Unlike many commercial grooveboxes with performance-oriented keys or pads, the Quadra prioritises a generative rather than a manual approach to beat making. Parameter controls on the Quadra are accessible through dial adjustments, allowing for live manipulation of mix, timbre, intensity, and other aspects. This mode of interaction can be likened to a more conducting-oriented approach rather than a traditional triggering approach to performance. With this combination of generative algorithms and live control, the Quadra encourages exploration and creativity that minimises the need for rhythmic performance skills.

Sequencers are a critical element of grooveboxes as they provide the patterns at the heart of their compositional capability. Most grooveboxes employ step sequences where users employ interface buttons to craft rhythmic and melodic patterns. Over time, sequence features have evolved and expanded, encompassing parameter sequencing, step sub-division, and probabilistic triggering. The Quadra takes an alternative approach by employing algorithmic sequence generation with variations managed via a global probabilistic trigger adjustment. This approach suits its limited interface and fosters access to those with minimal compositional experience.

8.3 A DIY Approach

During the decades in which the groovebox has evolved, the maker culture has also emerged as a thriving DIY electronics and fabrication movement. This cultural shift has resulted in broader access to various DIY technologies, positively impacting the community of music technologists (Chippewa, 2016). Within this community, a dedicated network of sound makers has emerged, using basic electronics to create a wide range of sonic devices. Among the notable creations are Michel Waisvisz's CrackleBox, a touch-controlled portable noise machine (Waisvisz, 2004), and Peter Vogel and Eirik Brandal's playable electronic sculptures. More recent non-conventional creations include the Monome Norms and Modern Sounds' Pluto. Furthermore, the maker culture has adopted digital fabrication techniques, such as 3D printing, laser cutting, and custom PCB manufacturing. These tools have enriched the landscape for DIY electronic instrument makers.

8.4 The Onboard Quadra

The OnBoard Quadra is a DIY groovebox designed and developed as an exercise in frugality and efficiency. Aligned with the culture of openness and accessibility, the Quadra utilises budget-friendly resources and open-source software. It uses an inexpensive ESP32 microcontroller and open-source software, including the M16 Audio Library written by the author (Brown, 2021). This straightforward device features a custom PCB and a simple interface comprising eight buttons, four dials, and a 16 LED ring, as illustrated in Figure 8.1. Its compact and handheld form factor is powered by a rechargeable battery, ensuring portability and ease of setup, making it an ideal companion for on-the-go music production.

8.4.1 Features

The Quadra supports real-time synthesis for four parts, including drums, bass, chords, and lead, with a combined capacity of up to 8 voices. Equipped with a multi-track sequencer featuring generative pattern algorithms, it facilitates dynamic and evolving compositions. There are performance controls over parameters, such as volume mix and tempo, using multiple buttons and dials. The Quadra has a headphone audio output to support portable music making. While the features of the Quadra may be surprising given its modest hardware, it is still quite a bare-bones groovebox barely having the features of the MC-303 and certainly lacks the 303's array of presets and samples. When compared to modern grooveboxes, like those from Elektron, the Quadra falls well short on features.

Nevertheless, it is important to note that the primary objective of the Quadra's design was not to maximise its features but rather to explore the extent to which the groovebox experience could be harnessed using minimal materials and a constrained interface. Throughout the development process, numerous decisions and trade-offs had to be carefully considered and the following sections outline some of these.

FIGURE 8.1   The OnBoard Quadra DIY groovebox.

8.4.2   Hardware

The central component of the Quadra is the ESP32 microcontroller, manufactured by Espressif. Despite Espressif's promotion of the ESP32 for its wifi and Bluetooth capabilities, the Quadra does not utilise these features (Espressif, 2023). Instead, the audio signal processing capabilities of the Quadra are derived from the speed and dual-core architecture of the Xtensa LX6 central processing unit (CPU) within the ESP32. Other microcontrollers that could be considered for a DIY groovebox include the Teensy 4 and the Daisy Seed, both of which are powered by the impressive ARM Cortex-M7 CPU and the latter has built-in high-fidelity DACs. While these microcontrollers offer more computational power compared to the ESP32, they also come at a higher price point, approximately three times that of the ESP32.

In the past, microcontroller audio output capabilities were limited due to constrained computational power and the absence of dedicated audio outputs. Consequently, DIY audio projects often featured raspy square wave tunes with abrupt organ-like envelopes. However, the landscape has evolved with the advent of modern microcontrollers, such as the ESP32, which increasingly support the Inter-IC Sound (I^{2}S) protocol for high-quality stereo sound output. To fully utilise this protocol, an additional decoder, typically found on a digital-to-analog converter (DAC) board, is connected to the microcontroller. This combination of audio protocol and converters has enabled sound-oriented projects, like the Quadra, to leverage the efficiencies of microcontroller mass production.

Furthermore, the Quadra features a custom-designed printed circuit board (PCB) on which those components are mounted below, and on top are user-interface components (Brown & Ferguson, 2021). Also mounted underneath the PCB are stereo amplifier boards and small loudspeakers, providing the groovebox with self-contained playability. The DAC board output can be connected to either headphones or an external speaker system.

The final hardware element is an optional laser-cut wooden case. This case serves a dual purpose: safeguarding the bottom-mounted components and enhancing the instrument's overall handleability. Laser-cut plywood is commonly used in DIY projects, and online tools, such as makercase.com, simplify the design of custom enclosures. For users who opt not to use the case, legs can be attached to the corners of the PCB to keep it stable for desktop usage.

8.4.3   User Interface

Like many electronic music instruments the Quadra is controlled by buttons and dials. Interface elements are kept to a minimum in keeping with the frugal design approach, and to offer a simple interface for novice users. The top row of four buttons trigger various states, such as starting and stopping the sequencer or pattern regeneration, while the four lower buttons select the active part. Combinations of these buttons enable alternative functions and can change the parameters that the four dials control. A 16-LED ring is the Quadra's display and, by default, shows the sequence steps for the selected part but can also reflect parameter values and the current sequence step when playing.

A challenge with such a limited interface is the prioritisation of controls, or what Cantrell called 'musical HCI' (Cantrell, 2017). For consistency, the button functions do not change, although button combinations can provide variations. For instance, while the top-left button serves as the playback control, holding the *shift* button and pressing *play* triggers a 'stutter' mode that repeats the current sequence step as a musical effect. The dials default to adjusting volume for each of the four parts but do change function depending upon which buttons are held. When the *shift* button is held the dials adjust four global controls: master volume; tempo; probability; and global filter cutoff. On the other hand, when the part/instrument buttons are held, the dials alter the sound for that part by manipulating filter cutoff (timbre), amplitude attack and release, and delay send amount, among other parameters. To address the design challenge, critical decisions were made about feature priorities based on an examination of choices made by previous groovebox designers and through reflection on performance experiments described in more detail below. These experiments focused on identifying the most essential live variations required by the musical genres for which grooveboxes are commonly utilised. The resulting interface design sought to prioritise access to the most crucial parameters.

8.4.4   Software

Most of the features that enable the Quadra's low-cost microelectronics to become an 'impossible' groovebox are based on the software. Several essential software foundations contribute to its capabilities. Two of these are supplied by Espressif, the ESP32's manufacturer. The first is efficient I²S audio support and the second is the Real Time Operating System (RTOS) that coordinates audio as a background activity and manages tasks across the dual cores. The third foundation is the M16 Audio Library developed by the author (Brown, 2021). This library is focused on efficient execution, even if that is at the expense of some audio fidelity. It draws on many techniques deployed in the Mozzi audio library designed for the even less powerful Arduino ProMicro MCU (Barrass, 2020). The M16 Audio Library seeks to optimise the Quadra's audio processing, taking into account the limitations of its cost-effective microelectronics.

All parts in the Quadra share a subtractive synthesis architecture, with some drum voices doing without a filter component which is the most computationally expensive element. All voices use just one oscillator, although the chord and lead voices use waveform blending techniques to provide a richer timbre. Amplitude envelopes are all percussive attack-release types further simplifying the sequencing systems to require onset-only triggers. There is a shared delay line for echo effects. Each voice remains monophonic up to the reverb, which operates in stereo.

The Quadra features a five-track sequencer, with dedicated tracks for lead and bass parts, and three tracks for the drum section, specifically for the bass drum, snare, and hi-hats. The chordal part utilises an arpeggiator, which employs two synthesis voices to enable harmonic overlap between successive notes. All sequences have 16 steps and the arpeggiator is free running, providing some useful pattern phasing for added musical interest.

The Quadra incorporates an algorithmic sequence generator, using Euclidian patterns for rhythms and random walk series for pitch contours and sets. Parameter values are tailored for each part. All steps in sequences and arpeggiators have probability triggering. The default global probability is 100%, ensuring all notes play while rests are silent. Lower probabilities skip notes, and higher probabilities introduce notes to replace rests. Holding the *preset* button regenerates patterns on the downbeat when playing. Holding *preset* and a *part* button triggers a new pattern specifically for that part. Holding the *mode* button shifts the sequence to a randomly selected new chord. These variations provide limited control over the specifics of sequences but can be used even by those without sophisticated compositional knowledge.

The Quadra's use of generative patterns and probabilistic triggering, rather than manual pattern composition, is seen as a deliberate feature rather than a limitation. This approach fosters musical co-creativity with algorithms (van den Oever et al., 2023), and is a type of 'intelligent' music pattern generation drawing inspiration from related research (Conklin, 2003); (Toussaint, 2005); (Vogl & Knees, 2017).

A final point on the software for the ESP32 is to note that while computing power for synthesis and pattern generation is impressive given the cost, the memory available for programs is modest. As a result, the Quadra's groovebox features are somewhat constrained not only by digital signal processing needs but also by the memory required to track data like dial positions, wavetable arrays, sequence values, and delay buffers. However, later versions of the ESP32, such as the S3 variant, improve this limitation with minimal cost impact. The tests described in the next section were performed on a version of the Quadra using the ESP32-S3.

8.5 Testing

To evaluate the capabilities of the Quadra as a groovebox, a series of three case study compositions were created. These were conducted as benchtop 'live' performances each inspired by a different exemplar of live groovebox music. Recordings of the exemplars and Quadra performances are available online (links provided below). While the Quadra was not designed to match the extensive features of commercial grooveboxes, these explorations were essential in confirming that fundamental aspects of groovebox performance practice could be achieved on the Quadra. It has been recognised that formal evaluation of live music-making systems is challenging, is often subjective, and can involve perspectives from the performer, composer, or audience (Stowell et al., 2009).

The methodology employed in this study encompassed several approaches, including the analysis of exemplar performances, autoethnographic reflection from the perspective of the author as designer, composer, and performer, and user testing involving practical activities with the device (Ellis et al., 2011). A review of the main compositional features of each exemplar was undertaken, including observed compositional features and performance interactions. A collated summary of these features is shown in the left column of Table 8.1.

After conducting the exemplar review, the Quadra features were systematically aligned with the identified characteristics to identify the most suitable approach for their implementation. In certain instances, this process necessitated the addition or extension of Quadra features, particularly when certain characteristics were deemed crucial and warranted inclusion. This mapping ensured that the Quadra's capabilities were tailored to capture the essence of the identified characteristics.

Next, a task-based approach to testing the Quadra's capabilities was undertaken (Wanderley & Orio, 2002). A compositional outline was created for each work that sought to highlight the

TABLE 8.1 A list of major musical and interface features in the exemplar works and the corresponding feature in the Quadra

Exemplar Characteristics	OnBoard Quadra Features
Synthesis or selection of sounds	Synthesis algorithms for each part
Sequence pattern composition/adjustment	Generative pattern creation
Transport and tempo control	Play/stop and Tempo control
Part (un)muting and volume adjustment	Part volume adjustment
Per-part filter cutoff (timbre) change	Per-part filter cutoff (timbre) control
Per-voice amplitude envelope adjustment	Per-part Attack and Release envelopes
Chord sequence following	Live chord change trigger
Adjusting delay and reverb effects	Adjustable delay and reverb amounts
Overall mix filter cutoff change	Overall mix filter cutoff change
Saving and recall of presets	No presets

identified characteristics whilst following the idiomatic form of the exemplar, often within a condensed timeframe. Each composition was performed several times with observations made and refinements applied at each iteration. Each performance was recorded for review, the final performance in each case is available online.

8.5.1 Case Study 1 – Trance

The first case study is a Trance-style performance, critical for assessing the Quadra's capabilities as a groovebox designed for electronic dance music genres. The exemplar was a video performance by Lukáš Kapek using a Roland MC-707, characterised as Trance style EDM (Kapek, 2023). This work relied on performative filter cutoff changes and quick part-level adjustments to introduce and mute parts.

The Quadra Trance Jam (Brown, 2023a) inspired by this exemplar starts with a heavily filtered drumbeat. The filter opens gradually, and other parts are introduced by bringing up their volume. The arpeggiated pattern's free-running nature deviates from the 16-step length of other parts, obscuring some repetitive elements common in Trance and EDM. Nevertheless, the interplay with the tightly repeating lead line creates intriguing rhythmic tension. The bass part is not as timbrally clear and distinct as might normally be expected in a trance style, but it's noted that a bass part was, perhaps uncharacteristically, not a strong feature in the exemplar. The piece proceeds with filter and level adjustments to various parts to change the texture over time. The Quadra dials enable both gradual and sudden parameter changes such that the lack of a part-mute feature does not significantly hinder the performance.

8.5.2 Case Study 2 – Down Tempo

The second example is a grungy down tempo music performance. The exemplar performance was by Oscillator Sync and titled #Jamuary2023-Day19, and performed on an Elektron Syntakt (Oscillator Sync, 2023). This performance relied on audio effects quite heavily, including the use of delay, reverb, and overdrive. While the Quadra's effects may not match those of more advanced commercial grooveboxes, its probabilistic sequencing kept the looping pattern engaging over time.

The Quadra Grungy Slow Jam (Brown, 2023b) based on this exemplar opens with solo bass tones. Although the AR envelopes on the Quadra do not sustain tones, slow-release tones provide a similar pedal point function. The lead and arpeggio parts are introduced next through volume increase. They have contrasting release characteristics as found in the exemplar track. The Quadra's limited polyphony affects the desired textural density for this genre. Heavy use of delay and reverb go some way to enriching the sound. Extreme use of the Quadra's reverb algorithm with percussive sounds reveals the simple bucket-brigade style algorithm. The drums enter to provide more momentum. The use of less than 100% probability becomes evident and does serve to add pattern variety and interest. The Quadra's soft clipping effect contributes to a lo-fi quality but with less grit than the exemplar track's distortion. The track proceeds with moments of drums dropping out and adjustments to harmonic parts. The drums re-enter for the track's concluding section with increases in the global probability thickening the texture toward the end.

8.5.3 Case Study 3 – Minimalism

The third example is an upbeat minimalist-style performance. The exemplar is a work titled Apparition by Hélène Vogelsinger (Vogelsinger, 2020). That work is performed on a Eurorack modular system rather than on a groovebox which, perhaps, only makes it even more challenging as a test case. The Quadra performance leverages its multi-part sequences and the ability to randomly generate new patterns and change chords on the fly to sustain interest.

The Quadra Minimalist Jam (Brown, 2023c) is inspired by this exemplar. The track opens with the lead and arpeggiator parts playing in a counterpoint pattern with short-duration envelopes providing a pointillistic texture. A delay effect adds complexity to this texture. The bass part then enters with a contrasting long release and sparse pattern. The global probability is set above 100% to introduce additional notes into the patterns so that each loop is different, yet recognisably similar. Live single-part pattern changes and sudden global chord shifts, reminiscent of minimalist works by Steve Reich, are applied. The Quadra drum track subtly contributes as a percussive element with short, enveloped sounds. The texture is elongated using delays, reverb, and by varying release times of the pitched sounds. The generated work is not as polyphonic as the exemplar and relies, instead, on probabilistic variations and delays to add detail. Audio quality, especially of the reverb algorithms, is noticeably lower than in the exemplar. The piece concludes with a manual volume fadeout of the drum and bass parts, followed by a fadeout of the lead and arpeggiator parts before the sequence stops. Despite not being explicitly designed for this style, the Quadra's capabilities prove well suited to minimalism.

8.6 Future Work

The development of the Quadra has demonstrated that an affordable modern microcontroller can effectively serve as a simple groovebox. While there are potential additional features that could be incorporated, the current interface is already nearing its capacity to maintain ease of use without excessive 'menu diving'. Commonly requested features that could be considered for future iterations include rhythmic swing, audio sync for collaboration with other devices, sound and sequence presets, MIDI connectivity, and audio sample support through an external SD card reader. Software adjustments for the first three are feasible with appropriate memory allocation, while MIDI and sample playback would necessitate hardware additions. Future research with the Quadra may concentrate on exploring music-making opportunities, such as developing tutorials, manuals, and workshops to support users in maximising the instrument's potential while embracing its limitations.

8.7 Conclusion

This chapter described the development of an accessible DIY groovebox, showcasing the potential of an inexpensive modern microcontroller to underpin an effective music-making device. The frugal design approach and algorithmic sequence generation enabled the extraction of rich and diverse electronic music from minimal resources and a limited interface.

Through a series of case study compositions inspired by exemplar performances, the Quadra's capabilities were put to the test, and it demonstrated its ability to capture the essence of various loop-based musical genres. While not designed to compete with more feature-rich commercial grooveboxes, the Quadra proved adept at providing engaging performances and enabling creative exploration.

The software foundations, including efficient I²S audio support, RTOS management of the dual-core CPU, and the M16 Audio Library, played a pivotal role in enhancing the Quadra's audio signal processing capabilities. Despite its modest memory resources, the Quadra managed to leverage these software elements to achieve multi-part synthesis and pattern generation. The minimalistic design choices and user-friendly interface cater to novice musicians, hopefully providing a versatile and enjoyable music-making experience. Although some features, such as MIDI connectivity and sample playback, may be desirable for future iterations, the Quadra's current features align with the ethos of simplicity and accessibility.

In summary, the OnBoard Quadra exemplifies the potential of combining frugal software foundations, an inexpensive microcontroller, and an efficient interface design to create a simple, but fun, DIY groovebox; which really should be impossible.

References

Akai MPC (2023, July 26). In Wikipedia. https://en.wikipedia.org/wiki/Akai_MPC

Barrass, T. (2020). *Mozzi*. https://sensorium.github.io/Mozzi/

Brown, A. R. (2021). *M16: Arduino Audio Synthesis Library for ESP8266 and ESP32*. https://github.com/algomusic/M16 (website), available online at https://github.com/algomusic/M16 [accessed December 2023].

Brown, A. R. (2023a) Quadra Trance Jam (audio), available online at https://soundcloud.com/thejmc/quadra-trance-jam [accessed December 2023].

Brown, A. R. (2023b) Quadra Grungy Slow Jam (audio), available online at https://soundcloud.com/thejmc/quadra-grungy-slow-jam [accessed December 2023].

Brown, A. R. (2023c) Quadra Minimalist Jam (audio), available online at https://soundcloud.com/thejmc/quadra-minimalist-jam [accessed December 2023].

Brown, A. R., & Ferguson, J. (2021). Enhancing DIY musical instruments with custom circuit boards. *Proceeding of UbiMus 2021*. 11th Workshop on Ubiquitous Music, Matosinhos, Portugal. 83–91.

Cantrell, J. (2017). Designing Intent: Defining Critical Meaning for NIME Practitioners. *Proceedings of the International Conference on New Interfaces for Musical Expression*, 169–173. www.nime.org/proceedings/2017/nime2017_paper0032.pdf

Chippewa, J. (2016). Sonic DIY: Repurposing the creative self—Editorial. *EContact!*, *18*(3). http://econtact.ca/18_3/editorial.html

Conklin, D. (2003). Music Generation from Statistical Models. *Proceedings of the AISB 2003 Symposium on Artificial Intelli- Gence and Creativity in the Arts and Sciences*, Aberystwyth, Wales. 30–35.

Davis, C. (2021). *A Feminist History of the Roland MC-505*. https://digitalcommons.pepperdine.edu/cgi/viewcontent.cgi?article=1648&context=scursas

Ellis, C., Adams, T. E., & Bochner, A. P. (2011). Autoethnography: An overview. *Historical Social Research/Historische Sozialforschung*, *36*(4), 273–290.

Espressif. 2023. "ESP32." Accessed Nov 25, 2023. www.espressif.com/en/products/socs/esp32.

Genius (2023). Genius Home Studio (website), available online at https://homestudio.genius.com/ [accessed December 2023].

Huttunen, M. (2021). *Tahti: A groovebox for the web browser* [Master's Degree Programme in Sound in New Media, Aalto University]. https://aaltodoc.aalto.fi/bitstream/handle/123456789/107033/master_Huttunen_Max_2021.pdf?sequence=2&isAllowed=y

Huttunen, M. (2023). Tahti Studio (website), available online at https://tahti.studio/ [accessed December 2023].

Kapek, L. (2023). Roland MC 707 – no samples, only internal synth sounds (website), available online at https://youtu.be/MN1iUwevhc0?si=FU62An_FKMrSE1R- [accessed December 2023].

Mämmi, J. (2023). *An Unlimited Instrument: Teaching Live Electronics and Creativity* [Master of Culture and Arts]. Metropolia University of Applied Sciences.

Oscillator Sync (2023). Jamuary2023 – Day 19 – Elektron Syntakt (video), availible online at https://youtu.be/lvlh4tHDAVQ [accessed December 2023].

Roland Corporation (2023). MC-303 Groovebox (website), available online at www.roland.com/global/products/mc-303/ [accessed December 2023].

Stowell, D., Robertson, A., Bryan-Kinns, N., & Plumbley, M. D. (2009). Evaluation of live human-computer music-making: Quantitative and qualitative approaches. *International Journal of Human-Computer Studies*, *67*(11), 960–975.

Taylor, T. D. (2001). *Strange Sounds: Music, Technology and Culture*. Routledge.

Théberge, P. (1997). *Any Sound You Can Imagine: Making Music / Consuming Technology*. Wesleyan University Press.

Tjora, A. H. (2009). The groove in the box: A technologically mediated inspiration in electronic dance music. *Popular Music, 28*(2), 161–177.

Toussaint, G. (2005). The Euclidean algorithm generates traditional musical rhythms. *Renaissance Banff: Mathematics, Music, Art, Culture, July*, 47–56.

van den Oever, M., Saunders, R., & Jordanous, A. (2023). Co-creativity between music producers and 'smart'versus 'naive'generative systems in a melody composition task. *Proceedings of the 14th International Conference on Computational Creativity*. ICCC 23, Waterloo, Canada.

Vogelsinger, H. (2020) Apparition (video), available online at https://youtu.be/eWDi_ppgkmk [accessed December 2023].

Vogl, R., & Knees, P. (2017). *An Intelligent Drum Machine for Electronic Dance Music Production and Performance*, In, *Proceedings of New Interfaces for Music Expression 2017*, 251–256. Copenhagen, Denmark: NIME.

Waisvisz, M. (2004). *The CrackleBox*. http://crackle.org/CrackleBox.htm

Wanderley, M., & Orio, N. (2002). Evaluation of input devices for musical expression: Borrowing tools from HCI. *Computer Music Journal, 26*(3), 62–76.

9

THE SOUNDSCAPE CUBE SYSTEM

A Method for the Construction of a Coherent Soundscape During Recording and Mixing

Tore Teigland

9.1 Introduction

The concept of the soundscape was first introduced by R. Murray Schafer in his 1977 book *Tuning of the World* (Krause, 2008). Schafer's definition of the term 'soundscape' includes all sounds from a particular environment that reaches the human ear. Since the field has evolved differently around the world, as well as across disciplines, there are a diversity of opinions about its definition and aims. Consequently, the use of the term has become idiosyncratic and ambiguous. ISO 12913-1:2014 aims to enable a broad international consensus on the definition of 'soundscape', to provide a common foundation for communication across disciplines and professions with an interest in soundscape stating the following three key definitions:

Sound sources: Sound generated by nature or human activity.

Acoustic environment: Sound at the receiver from all sound sources as modified by the environment, actual or simulated, outdoor or indoor, as experienced or in memory.

Soundscape: acoustic environment as perceived or experienced and/or understood by a person or people, in context.

(International Organization for Standardization [ISO], 2014, p. 1)

Essential to the interpretation of a soundscape is spatial hearing.

Spatial hearing is the capacity of the auditory system to interpret or exploit different spatial paths by which sounds may reach the head. . . . Using spatial hearing, the auditory system can determine the location of a sound source and 'unmask' sounds otherwise obscured by noise. It can also orient attention towards or away from a sound source. Spatial hearing is almost entirely underpinned by 'binaural' hearing: the comparison of the signal at one ear with the other ear. These comparisons are reflected in terms of differences between time and level and are termed, interaural time difference and interaural level difference . . . and are fundamental to nearly all spatial hearing.

(Culling and Akeroyd, 2012, p. 123)

DOI: 10.4324/9781003396710-9

To determine a location, the listener needs to decide both the direction and distance to the sound source. While the term direction is uncomplicated and needs no further elaboration, distance can be viewed as either egocentric or exocentric. Egocentric cues provide absolute distance information to the hearing system, while exocentric cues for a sound source provide relative distance cues compared to other sources (Loomis *et al.*, 1992). In audio production, distance is most commonly viewed, addressed and/or experienced as exocentric. In other words, the ability to hear that instruments in an ensemble are internally located at different distances from the listener, sometimes also referred to as 'ensemble depth' (Rumsey, Neher and Brookes, 2004).

Several interpretations of depth exist. Griesinger describes depth as "Sound sources appear to be behind the loudspeaker basis, and not in the loudspeakers themselves" (Griesinger, 2000, p. 9). Griesinger does not distinguish between the terms distance and depth. Rumsey on the other hand describes the difference between distance and depth in audio reproduction as follows:

Source distance is considered to be the perceived range between a listener and a reproduced source. Depth on the other hand is related to the sense of perspective in the reproduced scene as a whole and refers to the ability to perceive a scene that recedes from the listener, as opposed to a flat sound image.

(Rumsey, 2002, p. 660)

In this chapter, the term depth embraces all possible perceptions of depth including distance to source.

While reviewing literature to this paper, the author came across a study by Ekman and Berg on depth in sound recordings (Ekman and Berg, 2005). This study investigates how sound engineers approach and describe the concept of depth in a stereophonic recording by comparing interviews in light of research on the subject. From a total of 9 sound engineers; 2 subjects reported it easy to change perceived depth, 1 subject reported it neither easy nor hard, 2 meant the possibilities to succeed depended on the music, while a majority (5 subjects) reported it hard to change the perceived depth in recordings. Despite the fact that only 2 of the sound engineers in the study found it easy to change the perceived depth in sound recording, the interviews revealed that all 9 sound engineers were familiar with research on the subject. In the author's opinion, this demonstrates that putting theory into practice, or converting knowledge into skills, is not always an easy task.

In this chapter, room- and psycho-acoustic research and understanding is presented in context of a practical sound engineering perspective, aiming to close the gap between theory and practice. This article explores how key sound characteristics of sound sources interrelates with the acoustic environment with varying depth. The result is a system consisting of two cubes visualizing key sound interactions for the direct and reflected sounds, describing a didactic method for altering the perception of depth in stereo recordings using standard available tools only. The aim is to provide the reader with a tool that is easy to comprehend, building the confidence and competence to construct a coherent soundscape as envisioned during recording and mixing, which is accordingly perceived by the listener.

The soundscape cube system should not be mixed up with 'the sound-box': a visualization tool for analyzing sound sources:

A four-dimensional virtual space within which sounds can be located through: lateral placement within the stereo field; foreground and background placement due to volume and distortion; height according to sound vibration frequency; and time.

(Dockwray and Moore, 2010, p. 181)

9.2 Describing The System.

"Soundscape exists through human perception of the acoustic environment" (International Organization for Standardization [ISO], 2014, p. 1). Figure 9.1 demonstrates the basic concept of the system using a cube to visualize depth in a soundscape. The four musicians positioned within the cube are clearly localized differently both left to right and front to back in the depicted acoustic environment. Visualizing a scene that recedes from the listener, extending behind the speakers, creating the illusion of a three-dimensional space in a manner commonly found in 2D images. This basic visualization concept is applied to the following two discrete cubes which combined make up the soundscape cube system:

- The direct sound cube.
- The reflected sound cube.

While the direct sound cube addresses sound characteristics related to only the direct sound (i.e., sound sources), the reflected sound cube addresses sound characteristics related to only the reflected sound (i.e., reverberation from the acoustic environment). Each cube suggests a set of attributes to be customized according to the envisioned distances to the different sound sources making up the soundscape. In the following, each sound cube will be discussed separately.

9.2.1 The Direct Sound Cube

The height and width of the direct sound cube in Figure 9.2 decreases with depth – implying that high frequency content (HF), intensity and panning width for the direct sound also indicating with depth.

FIGURE 9.1   Illustrating the concept of the soundscape cube system visualizing depth.

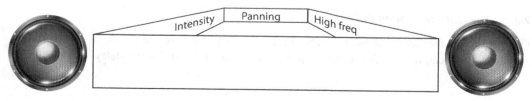

FIGURE 9.2   The direct sound cube – sound attributes in the direct sound on the consequence of depth.

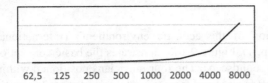

62,5   125   250   500   1000   2000   4000   8000

FIGURE 9.3   Typical increasing absorption nonlinearity in air vs frequency.

Source high frequency (HF)

In an open acoustic space, the direct sound will travel through air from the sound source to the listener. It is well established that the absorption of sound in air is frequency dependent, drastically increasing with higher frequencies (Everest and Pohlmann, 2015, p. 200), typically as illustrated in Figure 9.3. Consequently, increasing the distance will gradually alter the frequency response of the direct sound, attenuating higher frequencies more than lower frequencies. The result is that sounds with more high frequency content will be perceived as closer to the listener and vice versa (Levy and Butler, 1978).

Source intensity (level)

Differences in intensity have been investigated as a parameter to distinguish one sound from another. This research is based on the inverse square law which states that sound pressure changes by -6dB with every doubling of distance in a free field. As a result, the intensity of the direct sound will decrease with distance. Research has shown that a louder sound is perceived as closer to the listener and vice versa (Mershon and King, 1975; Blauert, 1997). Mershon and King found that "although a change in auditory intensity may be a good relative cue for auditory depth, it is ineffective as an absolute cue" (Mershon and King, 1975, p. 413).

Source panning

Changing the direction of a singular sound source through panning on its own will not affect the perception of distance. However, in context with reflections, arranging the sound sources panning similar to the common simulation of depth in 2D-imaginary by converging parallel lines as illustrated in the cube in Figure 9.2, will impart greater sense of distance-to-source. This forces the brain to automatically infer a 3-D context on the basis of such information being contained in the 2-D input of the retina (Shepard, 2001). In a study by Valente, Braasch and Myrbeck, they confirm a reverse relationship between distance-to-source and ensemble width. In other words, the perceived distance-to-source increases as panning width decreases. They state that in their scenario, the presence of visual cues is unlikely to account for the results, and that deviation between visual and auditory cues suggest they interrelate rather than one being underpinned by the other (Valente, Braasch and Myrbeck, 2012).

9.2.2   The Reflected Sound Cube

The height of the reflected sound cube in Figure 9.4 decreases with depth, while width increases – indicating that the direct to reflected sound ratio [D/R], pre-delay and decay for the reflected sound should decrease with depth, while stereo width should increase with depth. Furthermore, the additional triangular shape on the right side – that high frequency content in the reflected sound should also increase with depth.

D/R (direct to reflected sound ratio)

The perception of depth is dependent on the ratio between the direct and reflected sound. An increase in the reflected sound level (i.e., reverberation) compared to the direct sound leads to increased distance-to-source perception (Reichardt and Schmidt, 1966; Butler, Levy and Neff, 1980; Bronkhorst and Houtgast, 1999; Zahorik, 2002; Moon *et al.*, 2003; Griesinger, 2009). In an acoustic environment, the reflected sound energy remains roughly constant throughout (Moon *et al.*, 2003). As distance-to-source increases, the intensity of the direct sound will decrease according to the inverse square law as discussed in section 2.1. Consequently resulting in a decrease in D/R with distance as illustrated by the reflected sound cube in Figure 9.4.

Pre-delay

Pre-delay, sometimes referred to as the *initial time delay gap* (ITDG), specifies the time difference in milliseconds between the direct sound and the reflected sound from boundaries as they reach the listener. Looking at Figure 9.1, one can easily see that pre-delay will be shorter for the drummer (given the shorter distance from source to boundary to listener), than for the singer (given the longer distance from source to boundary to listener). Consequently, sounds intended to be perceived as close to the listener should have larger pre-delay, decreasing as distances to sources increase.

Decay

Research has shown that the decay time does not change significantly with distance in an enclosed acoustic space (Moon *et al.*, 2003). The same study also showed that late reflections had little impact on the perception of distance. By recording room impulse responses at various distances, combining them using early reflections from each distance, but late reflections only from the middle distance, listening tests participants were still able to judge distance with a satisfactory result. In effect, changing the decay time will therefore bare little influence on the perceived

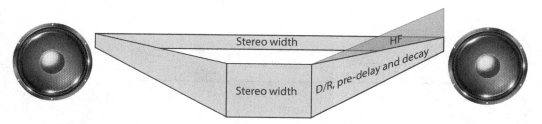

FIGURE 9.4 The reflected sound cube – sound attributes in the reflected sound on the consequence of depth.

distance-to-source. The soundscape cube system suggests that the decay time should decrease with distance-to-source. Doing so is a practical solution which can lead to better clarity/intelligibility by not having several long or similar reflections getting in the way of each other as the reverberation level at the same time increases with depth (see section 2.2 D/R). Shorter decay times also emphasize early reflections, which is critical to depth perception, as "we add early reflections to create the desired depth, and we add late reflections to bring out the hall" (Griesinger, 2000, p. 24).

HF (high frequency)

High frequency content in the reflected sound will increase with distance-to-source. To explain this, looking at Figure 9.1, the reflected sound for sources close to the listener (like the singer) will travel almost the entire length of the room before reflecting off the back wall. It then travels back the entire length of the room before reaching the listener. The increased distance travelled leads to more high frequency attenuation in the reflected sound (due to air absorption) compared to distant sources (like the drummer), from where the direct sound travels a shorter distance before reflecting of a wall and is projected towards the listener (almost half the distance). This is why, contrary to the direct sound, HF content in the reflected sound increases with distance-to-source.

Stereo width

Komiyama *et al.* showed that increased stereo width also increases auditory distance (Komiyama *et al.*, 2003). When lateral early reflections blend with the direct sound, the listener will experience the sum as one broadened image. This phenomenon is referred to as the apparent source width (ASW) and is commonly defined as the sound correlation between our ears within the first 80 ms of the onset sound (Lee, 2013). Increasing the stereo width of the reflected sound will therefore consequently also increase ASW. Lee showed that as the image is broadened, greater distance-to-source is achieved (Lee, 2013). Griesinger states that "the apparent width of the source increases and the position of the source moves toward the reflection" (Griesinger, 1997, p. 725), hereby effectively increasing distance-to-source perception. Furthermore, ASW is also recognized as an effective parameter to increase the sensation of spaciousness or spatial impression (Griesinger, 1997; Lee, 2013).

9.3 Using The System

Combining the two cubes; the direct sound cube and the reflected sound cube, into one three-dimensional figure as shown below, makes up the soundscape cube system.

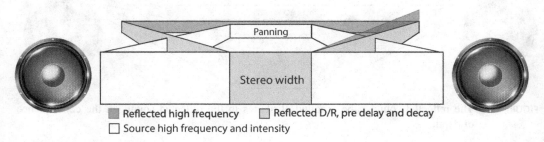

FIGURE 9.5 The soundscape cube system.

TABLE 9.1 Suggested values using the Soundscape Cube System on a vocalist positioned close to the listener and a drummer positioned far from the listener

Direct sound	Vocalist (close)	Drummer (distant)	Reflected sound	Vocalist (close)	Drummer (distant)
High frequency	Louder above 8–16 kHz	Softer above 8–12 kHz	**High frequency**	Hi cut, 1,5–2,5 kHz	Hi cut, 5–12 kHz
Intensity	Louder	Softer	**D/R**	High	Low
Panning	Center	Narrow <25–25>	**Pre-delay**	60–120 ms	0–20 ms
			Decay	1,3–2,5 s	0,5–1 s
			Stereo width	Narrow 30–50%	Wide 80–150%

By visualizing the location of different sound sources within the depicted three-dimensional space, relative adjustments to both the direct and reflected sound for the given parameters can be performed according to each sound source envisioned location.

9.3.1 A Practical Example

Table 9.1 shows suggested values when applying the system to the close positioned vocalist and more distant drummer in Figure 9.1.

Examining the three first columns with reference to the direct sound cube, high frequency content should be louder in the vocalist compared to the more distant drummer. Intensity should be louder for close sounds and softer for sounds as the distance increases. In Figure 9.1, both sources are horizontally centered in the room. Care should be taken to ensure that all individual elements making up the drum set are panned towards center, increasing the relative lateralization of reflections, to further support the impression of localization and depth.

Looking at the three last columns with reference to the reflected sound cube, high frequency content supporting the vocal should be considerably less compared to the high frequency reflections supporting the drummer. D/R should be higher (less reflected energy) for the vocalist and lower (more reflected energy) for the drummer. Pre-delay and decay should be larger for close sources and smaller for sources as the depth increases. Stereo width, on the other hand, should be smaller for close sounds and larger for distant sounds.

The suggested values used in Table 9.1 may well be considered typical. However, the ranges given are not delimited and other values should be considered and explored.

9.3.2 Which Tools To Use?

The Soundscape Cube System can be applied using standard available tools only.

For the direct sound

Intensity and panning are normally addressed with the fader and pan pot. High frequency content is easily modified using an equalizer, often choosing a hi shelving filter. High frequency content may also be adapted through microphone selection and placement.

For the reflected sound

In reverb units, pre-delay and decay is likely found as dedicated parameters within each unit. A way to adjust high frequency content is also often found, if not, adding an equalizer/filter in the signal chain before or after the reverb unit will work well. Stereo width can in some cases be found within each reverb unit but sadly often not. In such cases adding a mid-side processor after the reverb unit to adjust stereo width is a flexible and functional solution. The direct to reflected sound ratio (D/R) can be adjusted through reverb auxiliary send and/or return.

If using ambience microphone recordings for the reflected sound, the acoustic space sets the decay time and there is no good way to alter this besides adding decay time with a supplementary reverb. Pre-delay may to some extent be varied with microphone placement but can also be increased by adding a delay-unit to the ambience track or even increased or decreased by shifting the ambience track in a DAW timeline. By choosing and positioning ambience microphones, high frequency content can be altered or further modified using an equalizer/filter. Stereo width can be changed by adjusting the spacing between the ambience microphones used, or further controlled using a mid-side processor. Very slightly delaying one side of the ambience track will also change the stereo width and may in some cases produce very large stereo widths. The direct to reflected sound ratio (D/R) is simply adjusted by adjusting the fader of the recorded room-ambience. In a paper by Teigland and Jensen, the application of ambience microphones with regards to the direct sound is well covered (Teigland and Jensen, 2022).

Neither the direct sound nor the reflected sound can provide adequate localization separately (Griesinger, 2009). It is therefore important to understand that these parameters interconnect. Adjusting one parameter often affects the perception of another. As an example, a longer pre-delay will lead to perceived louder reverberation. Therefore, often adjusting just one or some of the parameters accordingly may yield the desired result.

9.4   Discussion

While reviewing numerous publications during the development of the soundscape cube system, statements such as: "Many aspects of spatial impression of reproduced audio are not fully understood yet" (Ekman and Berg, 2005, p. 1), or "several stimulus variables have been proposed as important factors in determining the apparent distances of sounds" (Mershon and King, 1975, p. 1), was frequently encountered. This chapter seeks to shed additional light on the complexity and in some cases the limitation of our understanding of depth perception, further examining the different parameters already reviewed.

For the direct sound cube described in section 2.1, intensity (loudness) and high frequency (spectral coloration) is well supported as depth cues (Mershon and King, 1975; Levy and Butler, 1978; Blauert, 1997; Everest and Pohlmann, 2015).

Source panning, on the other hand, is more ambiguous. Griesinger points out that

> careful listeners will notice that instruments to the far left or far right of a microphone array often seem closer to the listener than instruments in the center, even though the instruments in the center are closer to the microphone.
>
> *(Griesinger, 2001, p. 7)*

Valente, Braasch and Myrbeck discuss that this reverse relationship between distance-to-source and ensemble width might be influenced by the resulting increased lateralization of reflections (Valente, Braasch and Myrbeck, 2012). Griesinger supports this notion stating:

If we want a sound to be perceived as distant, we must add early reflections, and they must come from some other direction than the direction of the direct sound . . . Instruments far to the left tend to create reflections in the same channel as the direct sound. These reflections are masked. Early reflections in the right channel are usually not generated by sources at the left, so no depth perspective is created.

(Griesinger, 1997, p. 16)

Blauert *et al.* support this notion stating that "many room reflections are masked by the direct signal or by other reflections and are therefore inaudible" (Blauert, Mourjopoulos and Buchholz, 2001, p. 1). Furthermore, in a study by Rumsey *et al.*, "Unidimensional Simulation of the Spatial Attribute 'Ensemble Depth' for Training Purposes, Part 2" this phenomenon was accounted for while planning and conducting the study (Rumsey, Neher and Brookes, 2004). However, the same authors also pointed out one year earlier that lateral early reflections are *not* critical for the perceived ensemble depth (Neher, Brookes and Rumsey, 2003). Either way, panning sound sources increasingly towards center as distance-to-source increases is more efficient in presence of lateral reflections (stereo reverberation).

For the reflected sound cube described in 2.2, direct to reflected sound ratio as a depth cue is thoroughly described in numerous studies (Reichardt and Schmidt, 1966; Butler, Levy and Neff, 1980; Bronkhorst and Houtgast, 1999; Zahorik, 2002; Moon *et al.*, 2003; Griesinger, 2009). Reichardt and Schmidt reported that a 2 dB change in D/R is just noticeable by the listener (Reichardt and Schmidt, 1966). Later research by Zahorik expands this threshold to 5–6 dB, suggesting that D/R can only provide a coarse distance-to-source coding because of this, and that other cues probably provide a much finer resolution (Zahorik, 2002). Griesinger (2009) discusses this extensively, stating that

when the reverberation is below some threshold the singers are perceived as close to the listener – dramatically they occupy the same space. Just a tiny bit more and they leave the space of the listener and occupy a different space far away.

(Griesinger, 2009, p. 3)

Griesinger also points out that D/R interrelates with pre-delay:

As the pre-delay increases the effective loudness of the reverberation increases, allowing the engineer to increase the D/R while preserving the spaciousness. The clarity of the recording increases. Too much pre delay and the reverberation is perceived as an echo – this is clearly undesirable.

(Griesinger, 2009, p. 6)

This interrelation is central to the soundscape cube system. As nearby sources have larger pre-delays/ITDG, this concurrently supports lower reflection levels (e.g., increased D/R) making reflections still audible and avoiding masking by the direct sound. Very large pre-delay values, otherwise perceived as echo, might even still work if the level is lowered sufficiently. When the pre-delay decreases as sources move further away, louder reflection levels are required (e.g., decreased D/R) and greater distance-to-source perception is achieved. Although it makes sense that the reverberation of sounds perceived as close to the listener should have a larger pre-delay compared to distant sources, the author has not succeeded in finding any research where this interrelation is adequately described.

According to Bech, "the spectral energy above 2 kHz of individual reflections determines the degree of influence the reflection will have on the spatial aspects of the reproduced sound field" (Bech, 1998, p. 444). Much in line with this, the soundscape cube system suggests attenuating frequencies typical above 2 kHz for reflections from close sources (see Table 9.1), attenuating less as distance-to-source increases. As explained in section 2.2, this is due to increased air absorption in the longer sound paths for reflections from close sources compared to distant sources. However, the author has not succeeded in finding any studies where this relation between the amount of high frequency content in the reflected sound and perceived distance has been explicitly confirmed although the logic behind such a notion is clear.

It is well known that early reflections are dominant to late reflections in establishing distance to the sound source (Griesinger, 1997, 2009; Moon *et al.*, 2003), with Griesinger (2000) stating flatly that "we add early reflections to create the desired depth, and we add late reflections to bring out the hall" (p. 24). Moon et al., demonstrated that late reflections have little impact on the perception of distance (Moon *et al.*, 2003). According to Griesinger, changing the D/R has a far greater effect on the sound than changing the reverberation time by 30 percent or more (Griesinger, 2009). As the decay time of the reflected sound bares little influence on the perception of depth, it gives quite a bit of leeway. In this regard, the proposed concept of the soundscape cube system suggests that the decay time should decrease with distance-to-source. In other words, using long decay times to 'bring out the hall' for close sound sources, and shortening decay times with distance for distant sources, focusing on early reflections to increase distance-to-source. As pointed out in section 2.2, this is merely a practical solution which can lead to better clarity/intelligibility by not having several long or similar reflections getting in the way of each other, while still maintaining the desired depth perception.

A literature review by Kaplanis *et al.* on perception of reverberation in small rooms, describes width as the most commonly elicited attribute in reverberant and spatial sound fields (Kaplanis *et al.*, 2014). Width, or Apparent Source Width (ASW), is defined as the spatially- and temporally-fused auditory image of the original sound and early reflections (Ueda and Morimoto, 1995; Okano, Beranek and Hidaka, 1998). When lateral early reflections blend with the direct sound, the listener will experience the sum as one broadened image. There is a common understanding in the literature that perception of spaciousness is primarily related to lateral reflections (Kaplanis *et al.*, 2014). However, spaciousness is not exclusively associated to *early* lateral reflections (ASW), but also encompasses the impression of late reflections and Listener Envelopment (LEV) (Bradley and Soulodre, 1995).

9.5 Conclusion

The soundscape cube system describes a method for altering the perception of depth in stereo recordings. By combining two cubes, one for the direct sound and one for the reflected sound, the system visualizes how to adjust sound sources and reflections with changing depth.

According to Figure 9.5, the following sound adjustments should be performed with increased depth/distance-to-source:

The direct sound (source):

- Intensity should decrease.
- Hi frequency should decrease.
- Panning should be narrower.

The reflected sound (reverberation):

* Hi frequency should increase.
* Stereo width should increase.
* D/R should decrease (e.g., reverb increase).
* Pre delay should decrease.
* Decay should decrease.

Neither the direct sound nor the reflected sound can provide adequate localization separately (Griesinger, 2009). Thus, it is important to understand that these parameters interconnect. Adjusting one parameter often affects the perception of another. If used accordingly, good depth perception may also be achieved in mono.

9.6 Acknowledgement

The author wishes to thank Claus Sohn Andersen and Paal Erik Jensen for their critical reading and constructive feedback during the development of this article.

References

Bech, S. (1998) 'Spatial aspects of reproduced sound in small rooms', *Journal of the Acoustical Society of America*, 103(1), pp. 434–445. Available at: https://doi.org/10.1121/1.421098

Blauert, J. (1997) *Spatial Hearing: The Psychophysics of Human Sound Localization*. Cambridge, Massachusetts, London, England: MIT Press.

Blauert, J., Mourjopoulos, J. and Buchholz, J. (2001) 'Room Masking: Understanding and Modelling the Masking of Reflections in Rooms', in *Audio Engineering Society Convention 110*. Amsterdam, The Nederlands: Audio Engineering Society. Available at: www.aes.org/e-lib/browse.cfm?elib=9931

Bradley, J.S. and Soulodre, G.A. (1995) 'Objective measures of listener envelopment', *Journal of the Acoustical Society of America*, 98(5), pp. 2590–2597.

Bronkhorst, A.W. and Houtgast, T. (1999) 'Auditory distance perception in rooms', *Nature*, 397(6719), pp. 517–520.

Butler, R.A., Levy, E.T. and Neff, W.D. (1980) 'Apparent distance of sounds recorded in echoic and anechoic chambers.', *Journal of Experimental Psychology: Human Perception and Performance*, 6(4), pp. 745–750.

Culling, J.F. and Akeroyd, M.A. (2012) '123 Spatial Hearing', in Christopher J. Plack (ed) *Oxford Handbook of Auditory Science: Hearing*. Oxford University Press. pp. 123-144. Available at: https://doi.org/10.1093/oxfordhb/9780199233557.013.0006

Dockwray, R. and Moore, A.F. (2010) 'Configuring the sound-box 1965–1972', *Popular Music*, 29(2), pp. 181–197. Available at: https://doi.org/10.1017/S0261143010000024

Ekman, H. and Berg, J. (2005) 'The Three-dimensional Acoustic Environment as Depth Cue in Sound Recordings', in *Audio Engineering Society Convention 118*. Audio Engineering Society. Available at: www.aes.org/e-lib/browse.cfm?elib=13229

Everest, F.A. and Pohlmann, K.C. (2015) *Master Handbook of Acoustics*. Sixth edition. New York Chicago San Francisco: McGraw-Hill Education.

Griesinger, D. (1997) 'The psychoacoustics of apparent source width, spaciousness and envelopment in performance spaces', *Acta Acustica United with Acustica*, 83(4), pp. 721–731.

Griesinger, D. (2000) 'The Theory and Practice of Perceptual Modeling-How to Use Electronic Reverberation to Add Depth and Envelopment without Reducing Clarity', in *Preprint from the Nov. 2000 Tonmeister Conference in Hannover*. www.davidgriesinger.com/threedpm.pdf

Griesinger, D. (2001) 'The Psychoacoustics of Listening Area, Depth, and Envelopment in Surround Recordings, and their relationship to Microphone Technique', in *Audio Engineering Society*

Conference: 19th International Conference: Surround Sound – Techniques, Technology, and Perception. Available at: www.aes.org/e-lib/browse.cfm?elib=10076

Griesinger, D. (2009) 'The Importance of the Direct to Reverberant Ratio in the Perception of Distance, Localization, Clarity, and Envelopment', in *Audio Engineering Society Convention 126*. Audio Engineering Society. Available at: www.aes.org/e-lib/browse.cfm?elib=14920

International Organization for Standardization [ISO] (2014) 'ISO 12913-1: 2014 acoustics— Soundscape—part 1: definition and conceptual framework'. Available at: www.iso.org/obp/ui/ #iso:std:iso:12913:-1:ed-1:v1:en

Kaplanis, N. *et al.* (2014) 'Perception of Reverberation in Small Rooms: A Literature Study', in *Audio Engineering Society Conference: 55th International Conference: Spatial Audio*. Audio Engineering Society. https://aes2.org/publications/elibrary-page/?id=17348

Komiyama, S. *et al.* (2003) 'A loudspeaker-array to control sound image distance', *Acoustical Science and Technology*, 24(5), pp. 242–249.

Krause, B. (2008) 'Anatomy of the soundscape: Evolving perspectives', *Journal of the Audio Engineering Society*, 56(1/2), pp. 73–80.

Lee, H. (2013) 'Apparent Source Width and Listener Envelopment in Relation to Source-Listener Distance', in *Audio Engineering Society Conference: 52nd International Conference: Sound Field Control – Engineering and Perception*. Available at: www.aes.org/e-lib/browse.cfm?elib=16904

Levy, E.T. and Butler, R.A. (1978) 'Stimulus factors which influence the perceived externalization of sound presented through headphones', *Journal of Auditory Research*, 18(1), pp. 41–50.

Loomis, J.M. *et al.* (1992) 'Visual space perception and visually directed action', *Journal of Experimental Psychology: Human Perception and Performance*, 18(4), pp. 906–921.

Mershon, D.H. and King, L.E. (1975) 'Intensity and reverberation as factors in the auditory perception of egocentric distance', *Perception & Psychophysics*, 18(6), pp. 409–415. Available at: https://doi.org/ 10.3758/BF03204113

Moon, H. *et al.* (2003) 'The Effects of Early Decay Time on Auditory Depth in the Virtual Audio Environment', in *Audio Engineering Society Convention 115*. Audio Engineering Society. Available at: www.aes.org/e-lib/browse.cfm?elib=12437

Neher, T., Brookes, T. and Rumsey, F. (2003) 'Unidimensional Simulation of the Spatial Attribute "Ensemble Depth" for Training Purposes', in *Audio Engineering Society Conference: 24th International Conference: Multichannel Audio, The New Reality*. Available at: www.aes.org/e-lib/browse. cfm?elib=12271

Okano, T., Beranek, L.L. and Hidaka, T. (1998) 'Relations among interaural cross-correlation coefficient (IACC E), lateral fraction (LF E), and apparent source width (ASW) in concert halls', *Journal of the Acoustical Society of America*, 104(1), pp. 255–265.

Reichardt, v W. and Schmidt, W. (1966) 'Die hörbaren Stufen des Raumeindruckes bei Musik', *Acustica*, 17(3), pp. 175–179.

Rumsey, F. (2002) 'Spatial quality evaluation for reproduced sound: Terminology, meaning, and a scene-based paradigm', *Journal of the Audio Engineering Society*, 50(9), pp. 651–666.

Rumsey, F., Neher, T. and Brookes, T. (2004) 'Unidimensional Simulation of the Spatial Attribute? Ensemble Depth? for Training Purposes? Part 2: Creation and Validation of Reference Stimuli', in *Audio Engineering Society Convention 116*. Audio Engineering Society. https://aes2.org/publications/elibrary-page/?id=12635

Schafer, R.M. (1977) *The Tuning of the World*. New York: Knopf.

Shepard, R. (2001) 'Cognitive Psychology and Music', in P.R. Cook (ed.) *Music, Cognition, and Computerized Sound: An Introduction to Psychoacoustics*. Cambridge, Mass. London: MIT Press, pp. 21–35.

Teigland, T. and Jensen, P.E. (2022) 'Ambience Recording – The Influence of the Direct Sound When Combining Close and Ambience Microphones', in *Audio Engineering Society Convention 153*. Audio Engineering Society. Available at: www.aes.org/e-lib/browse.cfm?elib=21948

Ueda, K. and Morimoto, M. (1995) 'Estimation of auditory source width (ASW): I. ASW for two adjacent 1/3 octave band noises with equal hand level', *Journal of the Acoustical Society of Japan (E)*, 16(2), pp. 77–83.

Valente, D.L., Braasch, J. and Myrbeck, S.A. (2012) 'Comparing perceived auditory width to the visual image of a performing ensemble in contrasting bi-modal environments)', *Journal of the Acoustical Society of America*, 131(1), pp. 205–217. Available at: https://doi.org/10.1121/1.3662055

Zahorik, P. (2002) 'Direct-to-reverberant energy ratio sensitivity', *Journal of the Acoustical Society of America*, 112(5), pp. 2110–2117. Available at: https://doi.org/10.1121/1.1506692

10

DIGGING IN THE TAPES

Multitrack Archives as an Emerging Educational Resource

Paul Thompson, Toby Seay and Kirk McNally

10.1 Introduction

From the 1950s onwards, individual instruments and voices could be recorded separately onto their own track of a tape recorder. These multitrack tapes contain the recordings of individual musicians' instruments and their performances captured during a recording session. From an educational, musical, and historical perspective, multitracks can be seen to contain culturally significant fragments or moments of a recording session, which are additionally interesting if a multitrack contains an alternative take of an entire performance. Multitracks then can be thought of as seeing the different layers of a painting and how they all fit together. They may also include discarded layers, colours or other elements that weren't used in the final piece. This can help to show cultural practices, technologies, rooms, and spaces in a new light or some previously hidden insight into the intentions of the artist, the musicians, the engineers or the producer, and the production process more generally. Multitracks therefore emerge from a micro-system of creativity and using these materials can provide an opportunity to bring the music production process to life and uncover the often-hidden processes involved in creating a record.

The majority of multitrack audio holdings previously available, such as those found in the popular Cambridge Music Technology 'Mixing Secrets' website, have been edited and consolidated for ease of distribution and use, and generally don't include alternate takes, or out-takes, edit decision lists (EDL) or studio documentation such as tracksheets. This is changing however with new, more comprehensive multitrack sources becoming available, including commercial enterprises, such as the Sigma Sound Studios Collection at Drexel University (Seay, 2011) and the Open Multitrack Testbed (De Man et al., 2014). The EMI Music Canada Archive at the University of Calgary (UofC) in Canada is an emergent example that allows researchers to peer into the record company vaults. Importantly, in addition to the multitrack tapes of the recordings, this archive also holds demo tapes, song lyrics, concert planning documents, promotional material, cover art, correspondence between artists, management, producers, and executives. This presents new and exciting opportunities for researchers but also for audio educators to use primary source multitracks in the classroom to educate the next generation of audio engineers and producers.

The following chapter brings together previous work in this area (McNally, Seay, & Thompson, 2019; McNally, Scott, & Thompson, 2019; McNally, Seay, & Thompson, 2022) and draws on data gathered during a specific example of designing and delivering an intensive mixing module to

DOI: 10.4324/9781003396710-10

students from the Music Industry program at Drexel University using the resources gathered from the UofC Archives. A five-day intensive mixing project was designed to test our approach, respond to, and challenge traditional approaches that utilise multitrack audio materials and evaluate student learning and how it aligned with the module's intended learning outcomes. Students were asked to reflect and evaluate throughout the project, which helped to gain some real insight into how their approach changed over the course of the five days and what they learned as they did it. Beginning first with an introduction to the multitrack materials chosen from the UofC and then the structure and design of the module, the chapter then discusses the students' reflections on each stage of the process and concludes with an evaluation of the achieved learning outcomes within the categories of Creative Thinking, Lifelong Learning, and Inquiry and Analysis.

10.2 Digging In The Tapes: Beyond The Multitrack

In June 2019, a research visit to the EMI Music Canada Archive laid the foundation for delving into their collection of materials, which includes song lyrics, concert planning documents, promotional material, cover art, correspondence between artists, management, producers, and executives, and importantly demo recordings, in addition to the multitrack recordings. These additional resources also provide researchers a window into the creative choices made and the external factors influencing the artistic process outside the confines of the recording studio. Each record has its own unique story but usually the only stories available are the official stories from the marketing arm of the record company or official interviews with the artists used to promote the record. With this in mind, the research team searched through the archives for records that may have an interesting story that can be told through the archive's materials.

After three days at the archive, the research team happened upon the materials for the Grapes of Wrath, a Canadian alternative rock band that formed in 1983 but had their most notable commercial success in the late 1980s and early 1990s. The archives housed the multitracks and additional materials (such as internal EMI memos, faxes, invoices, press releases, and demos) for their 1991 album *These Days*, which was produced by notable British producer John Leckie. The 'story' of this record is interesting because it follows two previously commercially successful albums by the band, *Treehouse* (1987) and *Now and Again* (1989). *Treehouse* (1987) was the band's first album released by EMI Canada and it was produced by Tom Cochrane, frontperson for the rock band Red Rider. *Now and Again* (1989) achieved platinum certification in Canada, was produced by Anton Fier, the American drummer and producer, known for his work with The Lounge Lizards, The Feelies, Pere Ubu, and The Golden Palominos. *These Days*, the album used for this case study was produced by John Leckie, and again achieved platinum certification for the band in Canada. The album was recorded in Vancouver, British Columbia at Mushroom Studios and was mixed at Abbey Road Studio in London. The overall trajectory of the band as a popular and successful Canadian act was first leveraged in the US by EMI and then, following commercial disappointment there, the UK, which was targeted as a place to break the band into the market with a British producer.

The story of the record emerged by sifting through the materials, which included a memo from Terry McBride, the bands manager, to Deane Cameron and Tim Trombley at EMI Canada enquiring about securing Leckie as producer – interestingly, this memo pre-dates the making of the band's previous album, which indicates Leckie and the UK connection was a long-term goal. The search also revealed A&R notes concerning the quality, originality, and commercial potential of the band's album demos, in addition to the demo tapes themselves. Communications between Leckie's manager, Safta Jaffery and EMI Canada executives, which includes a listing of recent and current projects, further develops the story of why Leckie was chosen as producer and what he was expected to bring to the project with regards to both creative and commercial capital.

10.3 Hello From The Mooseland: Study Design

The module was delivered at Drexel University in Philadelphia by us, the researchers Kirk McNally, Toby Seay, and Paul Thompson, over a five-day period in November 2022. It involved nine first- and second-year undergraduate music production students. The Music Industry degree program at the Drexel University Westphal College of Media Arts & Design is a four-year undergraduate program that focuses on business, management, and music production. This degree program is consistently ranked as one of the best music business schools in the United States by various measures and students have access to state-of-the-art music production facilities and are taught by industry-tested faculty. Students take coursework in both business and creative areas of the music industry, with those who concentrate on music production taking advanced courses in recording, mixing, mastering, and arranging, and have the option to take advanced electives. All the students taking part in this module were music production concentration students. None had yet undertaken mixing-specific coursework, although all had digital audio workstation experience. They were recruited as volunteers by a call-for-participation to all Drexel music production students.

This module had two specific objectives, which were (1) to find useful applications of multitrack archival materials in audio education, (2) to challenge traditional methods of teaching the process of mixing. Storytelling was a way to connect both of these objectives. The story of the record was first constructed from the materials within the archives, and then the story was relayed to the students as they mixed and remixed the track. The students did not receive any technical instruction from the researchers, as is traditional in this type of activity, nor did the researchers offer any technical support in relation to the process. This helped to make the story, context, and the record's actors central to the project. Storytelling was used not only to remove any technical instruction but because it is a time-tested method of transferring beliefs, historical accounts, knowledge, skills, and practices (Abrahamson, 1998; Farrell & Nessel, 1982; McDonald, 2009). Rawatee Maharaj-Sharma argues that storytelling helps to

> locate the listener in a familiar context through the story. In a familiar context, the listener finds it easy to negotiate ideas presented through words, phrases, and even detailed accounts, to easily assign meanings and understandings by comparing and contrasting new ideas with existing ones. The feeling that they are being directly spoken to, whether it is through narratives or via digital media, takes the listener into the realm of the story with all its emotions and expectations (Dahlstrom, 2014). (Maharaj-Sharma, 2022: 3)

In this way, storytelling was used as a pedagogical approach to explore the ways in which storytelling can promote interest, familiarity, and active participation in the process. The module was structured in five steps and designed to reveal key information of the context and story of the record at different times throughout the students mixing project alongside the key individuals involved in the record-making process. The module also included in-person playback and discussion to allow students to discuss the project and clarify the requirements for each step. We asked the students to respond to new information and record their reactions by delivering a series of video reflections where they answered several questions that probed their thinking and decision-making processes as they worked their way through mixing and re-mixing the audio from the archives. This gave us as researchers some real insight into how their approach changed over the course of the five days and what they learned as they did it.

Step one involved enrolling students onto the Virtual Learning Environment (VLE), contacting them one week before the first in-person meeting to explain the mixing project, and give them access to the multitrack. They were asked to create a mix using the multitrack provided. The only information that was given at the time was the song title, the name of the band, the year the song was produced, the nationality of the album producer, scans of original track sheets and technical details regarding the origins of the digital files i.e. that it was transferred from analogue tape and an explanation of why the track count on the track sheets doesn't match the files distributed (sync reel bounces, etc.). Students were asked to deliver a short video diary before they started mixing so we could capture their initial response to the materials and their intentions, their ideas about certain parts of the arrangement, and how they planned to mix the song with reference to their existing knowledge of mixing techniques and in relation to reference material (i.e. other similar sounding songs).

Step two was the first in-person session with the students, which involved a playback and discussion of their mixes. This was followed by a presentation that introduced the 'story' of the record and involved a discussion of who the actors on the record were (musicians, engineer/producer, record executives, management, etc.), a historical framing of the era from which it was produced, which included heads of state, housing costs (to provide context for the album budget and producer's fee), and Billboard charts, etc. In the session, the students were asked to consider this information and redo their mix, imagining that they were the mixing engineer John Leckie.

In **Step three**, we asked students to capture their reflections and responses to the in-person session and then discuss how they were going to approach their next mix. We then asked students to produce a revised mix.

In **Step four**, we reconvened for a final in-person playback session of all the student mixes and played John Leckie's mix so they could compare and contrast against their own mixes.

Step five involved a final video diary, which provided students the opportunity to reflect upon the entire project and their response to this type of activity, what they liked, didn't like and what they perceived as the value of this type of exercise. An overview of the module timeline is shown in Table 10.1:

The in-person classes were videotaped using a single fixed-lens camera on top of the mixing desk's meter bridge to capture discussion and responses from the students. Both the in-person classes and the responses from the student video diaries were transcribed and thematically coded (Saldaña, 2015) to identify moments of surprise, reflection, evaluation, and learning.

TABLE 10.1 Timeline of 'Hello from the Mooseland' module

	STEP 1	STEP 2		STEP 3		STEP 4	STEP 5
	Video Diary #1	*Mix #1*	*Class #1*	*Video Diary #2*	*Mix #2*	*Class #2*	*Video Diary #3*
Task or Session	Pre-mix	First Mix	First Class Playback and Discussion	Post-Class Reflection	Revised Mix	Second Class Playback and Discussion	Final Reflection

Hello from the Mooseland, this module name, is borrowed from the phrase used by A&R representative Deane Cameron, who eventually became president of Capitol-EMI, in his salutations in letters to international representatives of Capitol-EMI around the world. The memos in the archives that use this salutation were sent to every Capitol-EMI branch office in the world, from India to Sweden, which in itself, reveals the network within which this single record was being produced in 1991. The salutation acts as a hook, but it also creates strong visual imagery and functions as both a stereotype and a sense of Canadian national identity.

10.3.1   Sound Reflections: Pre-Mix Video Diary #1

As mentioned above, students were asked to deliver their video diary *before* they started mixing and we asked them what they thought the intentions of the artist and the producer were in the studio and how those intentions are reflected in the multitrack. We asked them what ideas they might have about certain parts of the arrangement (i.e., how the introduction of instruments unfolds over time) and how they'll consider this in their mix. Finally, we asked how they plan to mix the song with reference to their existing knowledge of mixing techniques and in relation to reference material (i.e. other similar sounding songs), including a discussion of resources (both analog and digital), a timeline, and the ways in which they will evaluate the mix.

Excerpts from the first video diary submissions identified era-specific references (bands and sounds), while also discussing their creative approach to the mix, e.g. blending "old school drums and vocal" with "modern guitars" as shown below:

STUDENT 1: "I think it's a really cool song. I think that at least intentions wise it kind of feels like it's like a mix of like grunge and like '60s rock. Because I mean like, a lot of the guitars kind of sound like, grungy but the vocals remind me a lot of like, the Beatles. So, I'm thinking of trying to try to blend the mixing styles of both of those. To try to mix probably the drums and the vocals a little bit more old school and then try to do the guitars a little bit more modern. Try to blend them together a little bit."

STUDENT 2: "Overall, it reminded me of R.E.M. and that kind of style of alternative and new wave with a little bit of a '70s pop rock feel. I don't know if that makes sense but that's the vibe I got."

"I'm definitely a little intimidated by this project. I have very amateur experience mixing and this is probably the most tracks I've worked with all at once, which is a good thing."

"I'm not quite sure the direction I want to go with the song and I'll probably figure it out as I'm working on it, but I think that this is more suited to a kind of not small, but I don't think it requires big sound."

STUDENT 3: "My first thoughts when I listened to the track were that it was a little bit maybe alt rock, progressive rock, that kind of genre. And they definitely had that sound in the studio."

"I think I'm just going to do what I normally do. I'm going to take each track separately first and make it sound as good as I can individually sonically, make it a nice clean tone. And then I'm going to bring it all together in the end."

STUDENT 4: "My first thought is that it seems very like a big, very like '80s. I'm imagining using like a ton of reverb on like the snare drum and sort of making everything sound like sort of as big as possible, especially during the choruses."

"I'll probably start just by like doing all of the levels, making sure like you can hear everything properly. I'll probably go and do EQ then, like I don't even know if I need to do much compression. I think it all seems to be very consistent and like recorded well."

10.3.2 Sound Advice: Class #1

The in-person session with the students provided an opportunity to play back their mixes and for researchers and peers to offer some thoughts on their mixes and overall approach. This was followed by a presentation that helped to contextualise the archival materials and we began to tell the story of the record and presented the actors both on and around the production of the record (musicians, engineer/producer, record executives, management, etc.). With the record being older than our students it was important to provide a historical framing of the era so they could better understand and appreciate the record and materials from the archive. For instance, identifying the heads of state in 1991 helped develop political context, average housing cost provided a comparison for Leckie's producer fee and looking at the *Billboard* charts provided an opportunity for students to connect with other significant recordings from the era. We asked students to take all of this information onboard and re-visit their mix, imagining that they were the mixing engineer John Leckie. We also encouraged them to listen to Leckie's discography, so they could get a feel for his approach and resultant 'sound' – if that is possible!

10.3.3 Sound Reflections: Video Diary #2

We asked students to reflect on all the information they had been presented *before* starting their next mix. We asked students if they had revised what they thought were the intentions of the artist and the producer in the studio and, if so, how they can now hear/see those intentions reflected in the multitrack or somewhere else. We also asked them if they will revise the way in which they will approach the arrangement element of the mix (i.e. how the introduction of instruments unfolds over time). We asked them if they will revise how they plan to mix the song with reference to their existing knowledge of mixing techniques and in relation to the new information and other reference material. Finally, we asked what the most vital piece of information was that they thought helped them in revising their idea(s) or approach to mixing the song and how they will use this information in creating their next mix.

STUDENT 2: "I'm fresh out of our session, our Tuesday session, and I've just listened to the Stone Rose's album, The Second Coming, so everything's kind of fresh in my mind. I think after our session today, a lot of my assumptions were kind of affirmed. The kind of idea I had of the band in my head was definitely solidified. Where my approach, I think, is going to change a lot is creatively, now that I have a clearer picture of the intent behind the song and the intent that the producer had, especially taking into consideration the kind of industry pressure and the external pressure from the label and agents and stuff like that."

STUDENT 3: "I was really, really actually astounded by the amount of information you guys presented to us during the last meeting. And I honestly didn't even know about John Leckie before this assignment, but ever since that entire session where we just learned about the band, all the key players, and how it all came together and learning about John Leckie, I've just gone on the biggest rabbit hole through this man's discography ever."

STUDENT 4: "Okay, so I just got back from the first meeting for the study. I think the main thing I took away from it is learning who John Leckie is, learning about some of his material. I think I'll definitely look to that and think about that in regards to the way I mix the track. The album that I'm already the most familiar with is The Bends by Radiohead. I'll definitely be going back to that."

10.3.4   Sound Advice: Class #2

This in-class session again involved playback of students' revised mixes, mix critiques, and discussion about any new approaches or techniques they used. The class provided the researchers the opportunity to go beyond mix critique by contextualising their comments within the given storyline. It also served as an opportunity to explore how each student used the storyline to gain knowledge and perspective. In each case, the students' mixes had technically and aesthetically improved. This was the opportunity to explore why.

10.3.5   Sound Reflections: Video Diary #3

The third and final video reflection was designed to capture the students' response to the final mix, with an overall reflection and evaluation of the entire 'Hello from the Mooseland' mixing project. We asked them to include a reflection on the changes to the mix and their mix approach in response to the materials and how this will influence their mixing practice in future and an evaluation of the final mix and anything they'd revise if they had more time/resources (with specific reference to whether or not they thought they were able to meet the expectations of the artist and the producer in creating their final mix). We asked if they used all of the tracks in the multitrack and the ways in which this mixing project was different to previous projects they have undertaken. We asked them to mention the types of existing skills and knowledge of mixing they used in this project (if any) and what types of skills and knowledge of mixing they developed in this project (if any). We asked what the most vital parts of the module were that helped them in mixing this song and, finally, what they enjoyed most and least about the process and why.

The final reflections and evaluations captured some useful indicators of Creative Thinking, Inquiry and Analysis:

STUDENT 2: "Going forward, I think, definitely, um, just my mindset going in is going to change. I'm definitely going to be thinking more about what the purpose is, um, why I'm doing what I'm doing as a producer. Just being more thoughtful, I think, of what I'm doing in the mix and why, and sort of the external motivation for that. And I think just overall, the process of it, I think, the way I approach the mix, kind of focusing on the individual elements and comparing it to the mix as a whole, I think that process works pretty well."

STUDENT 4: "Thinking back to what I did differently between the first mix and the second mix, I think I was definitely thinking more like, okay, this needs to be like a hit single. This needs to chart and like, get on the airwaves. So, it needs to sound more poppy. So, there were definitely a lot of things I did differently. Yeah, I guess also just being willing to put more effects on the vocals and use reverbs and things, since I felt more like I was um, like a creative person sort of working with the band. Whereas the first mix, just getting the stems and being told to mix it I felt more like a like office worker or something to sort of like, just sort of guessing what the band might want and trying to do sort of like the least offensive, least um creative thing I could do."

10.4 Discussion

The purpose of this research project was to first find useful applications for multitrack archival materials within the context of audio education and, secondly, to challenge traditional methods of teaching the process of mixing that typically includes technically focused instruction. Storytelling was chosen to be central to constructing the story of the record from the archived materials, which was relayed to the students as they mixed and remixed the track. This work showed firstly that telling the story of the record, rather than simply asking them to undertake the process of mixing the multitrack, engaged students and made them feel part of the process. They were able to see themselves as balancing their own skills, knowledge, and taste with the requirements of the various actors involved in which:

STUDENT 2: "There's you know your own taste and creative ideas as a producer, but then there's also, you know, the intent of the band the musicians. There's the record label, all that kind of stuff. So, I think that taught me a lot and that kind of developed my understanding of your position as a producer and like, kind of what goes into that job."

In this way, the focus on the story allowed the multitrack materials to be used for learning about commercial record production on different scales in which 'Micro' involved studying the structure of the individual elements of the multitrack, 'Meso' involved viewing the relationship between the various materials and 'actors' in the process and, 'Macro' is where the archival objects point towards the larger story that includes socio-cultural elements such as commerce, fandom, genre, marketing, authenticity, and myriad other systematic agents. The micro level is often where educational systems are too often focused, especially when it comes to mixing different materials from different parts of the story and its actors allow for the examination of multiple scaled systems of creativity (Thompson, 2019, p. 239).

However, as shown in this case study, the entire creative system exists well beyond the participants in the recording studio, and examining this particular creative system, when scaled to the macro level, is afforded by holistic archival collections.

Secondly, this example also showed that engagement with the story of the record created new opportunities to learn about different processes and technologies such as SMPTE and multitrack tape and the ways in which they were used at the time. This prompted problem-solving and critical engagement with the ways in which the multitrack materials had been created as in the example: "a lot of the problems that we've been running into, or the nomenclature of it all, stuff that we just never dealt with, because it's never kind of been part of our mix environment (STUDENT 1)." In this way, students were introduced to different ways of working through the story, which was effective in facilitating learning as in previous studies using storytelling in the classroom (e.g. Cooper et al., 1992).

Thirdly, while the use of storytelling in more formal educational contexts is not novel, for example it has been used as a way to teach science in the classroom (e.g. Maharaj-Sharma, 2022), this case study is the first documented example of using archival materials to construct a story and use it as a way to facilitate the development of students' mixing practice within an audio education context. Instead of the traditional student-professor relationship (Wilson et al., 2010), the focus was placed on the students' own development, which was captured through personal reflection in response to the record's story. In this way, the project stimulated self-reflection, critical thinking, and intrinsic exploration and, importantly, helped to steer students towards their own conclusions. The context is important here too because the very first mix assignment involved almost no parameters and no grades were assigned at the end. This meant that experimentation was actively encouraged and students were able to reflect on the things that did or didn't work.

Finally, removing technical instruction in place of storytelling also helped to characterise mixing as a contextual, socio-cultural process that is part of the story and the context of a recording, rather than an individually focussed technical exercise. This is significant because it interconnects with the first point here where students can feel part of a process that goes beyond the multitrack materials.

10.5 Conclusions

This project has shown how multitrack archival materials can be collected and curated in order to tell the story of a record, which helps to create new associations between the various artefacts, and allow meta-narratives (or intra-narratives) to be constructed through macro, meso and micro examination. This is only possible when there are additional archival materials beyond the multitrack where they can be assembled to help retrace the story of the record. Storytelling was used in place of traditional methods of teaching the process of mixing, which usually involves technically focused instruction, and provided a way of relaying aspects of the production of the record as the students mixed and remixed the track. By working directly with the materials, the students become part of the record's story and they can see more clearly how they are part of the record's creative systems simultaneously. Importantly, this project has highlighted how the use of archival materials can:

1. Bring students closer to a mixing project;
2. Allow students to see themselves as part of a creative system, rather than an individually focused assignment;
3. Help students to connect their approach and intent to a contextualised broader system of cultural production;
4. Encourage students to listen to the work of other engineers and producers in a critical way (in this case, John Leckie);
5. Introduce students to different musical styles/eras/approaches;
6. Encourage Creative Thinking, Lifelong Learning, and Inquiry and Analysis.

This research also provides a useful example of how materials within these archives can be mined, curated, and used and serves as a blueprint for other researchers and educators who may wish to use archival materials and storytelling in teaching mixing. It also functions as a stepping-stone for further work in developing a curated package of archival materials for educators to access and deliver as part of their own programmes. Even beyond the recording studio or classroom, this work has implications for the use of non-traditional approaches in audio education and further work is needed to test its efficacy in teaching different aspects of studio production.

Acknowledgements

The authors wish to thank the following institutions and offices for their support of this research:

- Social Sciences and Humanities Research Council of Canada
- Archives and Special Collections, University of Calgary
- Drexel University's Antoinette Westphal College of Media Arts & Design – Rankin Scholar in Residence
- Drexel University Office of Global Engagement
- Leeds Beckett University

References

Abrahamson, C. E. (1998). Storytelling as a pedagogical tool in higher education. *Education*, 118(3), pp. 440–451.

Cooper, R. Collins & Saxby, M. (1992). *The Power of Story*. Melbourne, Australia: Macmillan Education.

Dahlstrom, M. F. (2014). Using narratives and storytelling to communicate science with non expert audiences. *Proceedings of the National Academy of Sciences USA*, 111 (Supplement 4), pp. 13614–13620.

De Man, B., Mora-Mcginity, M., Fazekas, G., & Reiss, J. D. (2014, October). The open multitrack testbed. In *Audio Engineering Society Convention 137*. Audio Engineering Society. https://aes2.org/publications/elibrary-page/?id=17400; www.aes.org/publications/conventions/?num=137

Farrell, K., & Nessel, D. D. (1982). *Effects of Storytelling: An Ancient Art for Modern Classrooms*. San Francisco: Word Weaving.

Maharaj-Sharma, R. (2022). *Using Storytelling to Teach a Topic in Physics*. Education Inquiry, pp. 1–20.

McDonald, J. K. (2009). Imaginative instruction: What master storytellers can teach instructional designers? *Educational Media International*, 46(2), pp. 111–122.

McNally, K., Scott, K., & Thompson, P. (2019). Towards a Pedagogy of Multitrack Audio Resources for Sound Recording Education. In *Proceedings of AES 147 Convention*, New York, October 16–19, 2019. www.aes.org/tmpFiles/elib/20240729/20670.pdf

McNally, K., Seay, T., & Thompson, P. (2019). What the Masters Teach Us: Multitrack Audio Archives and Higher Education. *The Bloomsbury Handbook of Popular Music Education: Perspectives and Practices*, Moir, Z., Powell, B., & Dylan Smith, G., editors. Bloomsbury Publishing, London UK. pp. 113–126.

McNally, K., Seay, T., & Thompson, P. (2022). Multiple Takes: Multitrack Audio as a Musical, Cultural, and Historical Resource. Presentation as part of the Innovation in Music 2022 Conference, Royal College of Music, Stockholm, Sweden.

Saldaña, J. (2015). *The Coding Manual for Qualitative Researchers*. Thousand Oaks, CA: Sage.

Seay, T. (2011). Primary sources in music production research and education: Using the Drexel University Audio Archives as an institutional model. *Journal on the Art of Record Production*, 5.

Thompson, P. (2019). *Creativity in the Recording Studio: Alterative Takes*. Basingstoke: Palgrave Macmillan.

Wilson, J. H., Ryan, R., & Pugh, J. L. (2010). Professor-student rapport scale predicts student outcomes. *Teaching of Psychology*, 37(4), pp. 246–251.

Discography

The Grapes of Wrath (1987), [CD] *Treehouse*, Canada: Capitol Records-EMI Canada.

The Grapes of Wrath (1989), [CD] *Now and Again*, Canada: Capitol Records-EMI Canada.

The Grapes of Wrath (1991), [CD] *These Days*, Canada: Capitol Records-EMI Canada.

11

GATEKEEPING IN THE AUDIO MASTERING INDUSTRY

Russ Hepworth-Sawyer

11.1   Introduction

This chapter begins with an introductory discussion around the principles of audio mastering, and most importantly how it fits within the music industry context. Gatekeeping is then considered from a theoretical standpoint and how this is applied to other industries, before exploring a music or record industry model pre-1999, the point at which the industry is disrupted by Napster (Page, 2021). As a backdrop, in this case the record industry, the chapter demonstrates the needs and function of a gatekeeping system. Post-1999, upon the emergence of Napster, gatekeeping shifts and adjusts based on disruption to the market through technology and how this crisis was predicted to happen, which is discussed later. With such significant disruption, the financial structures, or the ecosystem, plus the supply chains are altered. Next, we discuss audio mastering's specific gatekeeping and how this has adjusted over time and has been adjusting the way in which mastering has earned its income and those involved in the practice. Dolby Atmos and, more widely termed, spatial audio, might be the next stage in gatekeeping to aid audio mastering businesses to retain their income, and purpose, going forward. The chapter concludes with a discussion highlighting the dichotomies of both democratization and gatekeeping within this small section of the wider music industry.

11.2   Audio Mastering Context

As the last stage in the production process, audio mastering is a vital creative step to ensure quality control, and furthermore the widest possible acceptable translation to audio systems across the world now, and into the future. Audio mastering involves receiving balanced files from a mix engineer, or perhaps a self-producing band, and either processing this to improve the overall tonal and dynamic outcome as a standalone release or matching this work with other tracks to form a playlist, EP, or album (Cousins & Hepworth-Sawyer, 2013).

As a field, audio mastering is technical in so much that it requires many of the same forms of equipment found in a recording studio. Equipment such as loudspeakers (called monitors),

DOI: 10.4324/9781003396710-11

dynamics processors, equalizers and perhaps some form of router (or mastering console), are all found in the mastering studio (Hinksman, 2020). Also, however, many mastering studios will house items such as lathes for cutting acetate for vinyl manufacture. These items are, naturally, expensive and require significant financial investment.

The creative discipline of becoming a mastering engineer is also to be discussed within the sphere of this chapter's topic. The mastering engineer must not only decide to become a mastering engineer but also be in a position to become competent and finesse their skills to handle a wide range of musical source material. This human investment is considerable and can be considered as the value, or 'capital' of the mastering engineer. My research on this, yet to be published, suggests that the period of time necessary to become a competent professional may be well in excess of an architects' degree course of seven years. The financial and personal investments outlined above act as gatekeepers to participating at the professional level of audio mastering. However, there are additional gatekeeping elements that will be investigated throughout this chapter.

11.3   What Is Gatekeeping?

Gatekeeping as a theory is widely attributed within the field of media studies and relates to the necessity to separate out, or select, items for publication (or broadcast) from a wider selection of content. Naturally we can conceive this as the newspaper editor "culling and crafting countless bits of information into the limited number of messages that reach people each day" (Shoemaker et al., 2009). Further exploring the concept of the term 'gatekeeper', even in common parlance, we might consider the role of a PA to a busy CEO of a firm whereby calls, meeting requests, letters and any interruption are sidelined without the knowledge of the person the gatekeeper serves.

Janssen & Verboord (2015) consider the theory of gatekeeping the likes of "publishers, film studios, gallery owners, critics or reviewers" being defined as "those involved in the mediation between production" and the consumer (Janssen & Verboord, 2015). This naturally links to the work of Bourdieu who said that "there is an economy of cultural goods, but it has a specific logic" (Bourdieu, 1984). Linking gatekeeping to the increase of artistic value is key here and as such gatekeeping is, therefore, not a new concept in the music industry. Sanders et al. (2022) highlight the natural gatekeeping that has been the dimension between the 'sound sellers' and any intermediary between them and the buying (or now, merely listening) public. Sanders et al. also express that these gatekeepers are 'cultural intermediaries' closing a loop between 'creativity and commerce' (Sanders et al., 2022).

Maguire & Matthews (2010) also quote Bourdieu (1984) and consider that industry gatekeepers can be considered "producers of symbolic goods and services". Sanders et al. (2022) mold this to highlight those gatekeepers "mediate between the production of cultural goods and the production of consumer taste", and thus "lend legitimacy to a curated set of cultural products and thus build economic value". This is an apt explanation for the existence of gatekeeping being the modus operandi of the historic music industry, and why curation ultimately maintained value.

Gatekeeping, as it is seen through this lens, places responsibility upon the record label professional to ensure the financial viability of the creative product, and thus the return on investment made. This understanding is historically quite accurate and a good barometer for the cultivation of creative music production and talent intended to generate a financial profit. Negus (2011) directly associates the Bourdieu idea of 'cultural intermediaries' when considering gatekeeping in the record industry (Negus, 2011).

11.3.1 Gatekeeping In The Record Industry

In this section, the way in which the industries are described is important in terms of distinction. The record industry refers to the sale of recorded media (and more modern streaming systems). Meanwhile, the recording industry refers to the wider industry including the supply chain attached to the sale of music, in this case including the recording and mastering studios. The music industry is also referred to and is concerned with the widest encapsulation of the industry and might include touring, music instrument sales, music technology and beyond.

The music industry, in its widest consideration has always been robust in terms of business model. will.i.am, the producer and performer behind the Black-Eyed Peas during a *Wall Street Journal* video explains his interpretation of the construct of the music industry. In the video, interviewer Christopher John Farley attempts to extract some answers to the changing industry and funding concerns around streaming. will.i.am explains, "music was only ever there to sell something else". Whilst profound as a statement, will.i.am goes on to explain that sheet music, where the music industry arguably began in its widest form, was there to sell pianos, violins and 'tubas'. People would amass in front rooms around the world and perform to each other (will.i.am, 2013).

Indeed, the sheet music industry remains with us to this very day in the term 'publishing' in recorded music. The initial industry was made up of this construct with the songwriters being the stars. However, as soon as the reasonable method to reproduce and distribute reliably recorded sound (in the form of records), the record industry was born. However, if we concentrate upon the native business of selling records, especially in the period prior to 1999, we can identify a clear example of gatekeeping in action, and to successful effect.

Prior to the aforementioned disruption caused by Napster in 1999, the record industry was reasonably healthy and had a traditional, yet robust, business model. The model had remained mostly stable and, in many cases, grew since the introduction of widespread access to the record from the 1930s (Page, 2021). In fact, Patrik Wikstrom wrote the following:

> In 1999 the global music industry had experienced a period of growth that had lasted for almost a quarter of a century . . . At the end of the nineties . . . Shawn Fanning would ignite the turbulent process that would eventually undermine the foundations of the industry. (Wikstrom, 2014)

One only needs to compare the number of sales required to clearly reach the no.1 slot on the UK charts in what could be called the 'heyday' compared to 1999 to observe relative stability. The best-selling single sales ensuring a no.1 slot in the UK charts in 1963 with the Beatles' 'She Loves You' reached 1.93 million, and 1999's Britney Spears' 'Baby, One More Time' reached sales of 1.45 million (*MusicWeek* 2000). Tatu's UK no.1 'All The Things She Said' only required 90,000 sales to reach no.1 in 2003. Ultimately it only sold 764,000 sales by 2022 in comparison (Official Charts Company, 2023a, 2023b).

This 'stable' business model that not only revolved around the signing and management of artists but also managing assets included: recording studios the artists recorded in; pressing plants that pressed the vinyl; and beyond this, the distribution companies too (the wider music industry). In further cases, such as Virgin (UK), the industry also owned their own record stores too. The labels' model was thus that they would not only find artists to consider, then sign them, but they would also engage in building their careers through artist development and a robust marketing strategy. One might consider this a holistic approach.

Exploring this business model further, the marketing strategy towards artists would require them to often tour extensively to advertise their music to a wider audience, with the sole intention to drive sales in their pre-recorded music – the album, or indeed single(s). This could be a loss leader given the sheer profit enabled from the sale of recordings.

Oleinik's (2015) work on gatekeeping developed what he called 'the triad' between *gatekeepers, established actors,* and *would-be actors.* In Figure 11.1, this model has been mapped to the pre-1999 record industry. The gatekeeper is identified as the record label, where curation is key. Any record label making the investment required on a new act was only doing so to gain a return on investment (ROI). Therefore, the character of the artist, their musical qualities, their demeanor and so on, would all be considered deeply before committing to any large sum of money. The diagram considers that there is a large pool of equally talented musicians to choose from who are denied the same exposure and opportunities as those who are signed. From the musicians' perspective, the gatekeepers were difficult to access, let alone convince. However, from the label's perspective, there is only so much money to invest, and a high requirement to achieve a ROI. From the consumer's perspective, however, whether consciously considered or not, the music on offer was curated by the record label professionals. Only the best, or intended, acts made it to the radio playlists and record stores, by virtue of the business model – in those days based on record and cassette sales (prior to the CD's eruption into the marketplace in 1983).

The established actors are the artists who have been signed, or 'curated' to appear on the record store shelves and within your morning radio consumption on the way to work. In the image, I've considered this in tandem with the record label itself, tying the gatekeepers to the entry barrier of the label. The would-be actors are the aforementioned, equally talented musicians, alongside those training, who are seeking to become a signed artist and gain the prominence, marketing and exposure that only a record label could provide.

Relating this to the audio mastering business within the same era (the 1980s in this case), the established actors are the audio mastering engineers gainfully employed, and the would-be actors are those who might (if they knew about mastering) aspire to be gainfully employed mastering engineers. The restrictive gatekeeping element in this period would have been the fact that few people knew audio mastering existed and would know far less about the finer details.

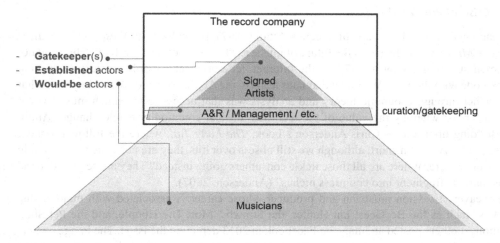

FIGURE 11.1   Traditional music industry gatekeeping before 1999.

11.3.2 What Happens If Gatekeeping Shifts Or Is Removed?

As introduced earlier, Napster launched in 1999 and began a chain reaction which was to change the record industry completely and forever (Page, 2021). The access to this music that Napster provided was permitted in a manner where those who made the music (the established actors) and invested in the music (the gatekeepers) received none of the dues expected. This development was, of course, the beginning. What must be highlighted is that the move to legal streaming with the likes of Spotify, Apple Music and Tidal, amongst others, was not created from the music industry itself. Hurjan et al. (2020) suggest that "many digital businesses succeeded to adapt and utilize new technologies [whilst] many traditional businesses failed to seize such opportunities to adapt and survive the market" (Hurjan et al., 2020).

This exposed the ultimate fragility of the system and business model that supported the record companies and by proxy, the wider music industry. With such disruption to the record labels, the ricochets felt were deep, and in some cases terminal. In this context, it encapsulated all the smaller businesses – what we might consider the supply chain that supported the functions of the once-resourceful record company operation. The less obvious contenders affected by a lack of dominance in the physical distribution of legal music would be the pressing plants, distributors, printers, typesetters, and graphic designers, just to name a few. The more obvious contenders would be the recording studios who no longer can charge the same day rates they once did for a recording session, as the labels could not pay what they once did on their new investments. This ricochet naturally extended to the mastering studios (which the recording studio output must be processed before release). In other words, the record industry's product would have once financed the record industry's wider operation.

However, in contemporary times, the most popular distributors (streaming companies) are independently owned and could be classed as computer data hosting companies. If one looks at Spotify on this level, how is it any different to a paid-for video streaming service, or a photo library service? The raison d'etre of these computer companies is not the record label, the musician and certainly not the wider music industry the label's operations once served. In fact, this transition to streaming has enabled a prophecy to come true.

11.3.3 Crisis of Proliferation

Many reading this will be aware of Jacques Attali's 1977 pivotal work, *Noise: The Political Economy of Music*, where he terms the future of music as "the crisis of proliferation". He could not have known about streaming in 1977, or the Internet, let alone that portable music consumption would explode with the Sony Walkman that later leads to the Apple iPod and so on. However, what he did predict through economic theory and analysis was that access to too much music would result in a 'crisis' whereby the value of the recorded musical commodity would change (Attali, 1977). Building upon this is Chris Anderson's book, *The Long Tail*, where the following quote supports Attali's view: "In short, although we still obsess over hits, they are not quite the economic force they once were. Where are all those fickle consumers going instead? They are scattered to the winds as markets fragment into countless niches" (Anderson, 2007).

Blue Weaver, a session musician and producer whose career is associated with major A-list 1970s' acts, such as the BeeGees, Ian Hunter, the Strawbs, Mott The Hoople, and the Pet Shop Boys, amongst others posted an image on Facebook in 2013 depicting the issue. The image depicts a box of A4 paper, at least 20 cm tall. The paper represented the printouts of his royalty statement for that year. Beside the box of paper is an expensive bottle of French red wine (Figure 11.2). Weaver's caption accompanying the image says, and I paraphrase, alongside a personal message,

FIGURE 11.2   Blue Weaver's image from a Facebook Post regarding dwindling royalty payments.

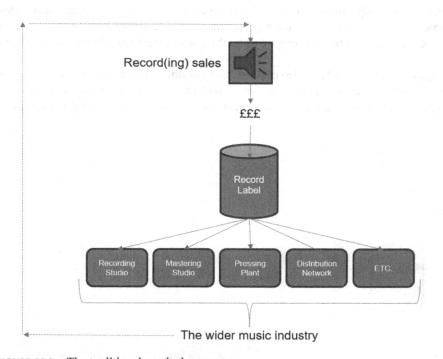

FIGURE 11.3   The traditional musical economy.

that – this image reflects the state of things at [my record label]. "My royalty statement contains all that paper and not enough income to buy the bottle of wine".

The above example explores the issues around income and investment, not just within the record label itself, but within the wider music industry that also suffers (Figure 11.3). The model below details the principle of the record label effectively self-investing in the wider music industry. Within this model, whilst under the influence of gatekeepers – an essential quality perhaps, according to Attali to ensure value – the music industry continued to be able to financially support the 'supply chain' through fees commensurate with the trade.

However, whilst many might welcome-in the new technology and ubiquitous access to all music, all of the time, the financial structure has altered. The computer sector companies, or 'Digital Service Providers', as Towse called them, are now streaming the music and have shareholders that require payment, as illustrated in Figure 11.4. As the digital service provider is really only a data company, where "some . . . do not even specialize in music as part of their wider business interests", there is no requirement to invest in the musical economy that supplies it (Towse, 2020).

Over a period of time, the fees that studios have been able to charge have reduced, as the labels provide smaller advances and budgets to generate the next product. This pressure has meant the loss of scores of studios over the period since 2000. Pricing is one of many aspects of this structural change and the inability for studio prices to maintain their value over the decades since. I recall the dry hire rate for equivalent gatekeeping work (CD mastering and manufacturing preparation) in 1994 at the studio I worked at being £45 per hour. According to an assortment of websites that provide a calculation of financial value across the same period (see below), 1994's £45 would be approximately £90 in 2023. As a quick analysis, I currently charge £60 per track, crudely meaning I could be being underpaid £30 for similar work. This change has forced most mastering engineers to set up as a sole trader, on their residential property as this reduces the overheads considerably, thus absorbing losses in earnings to keep trading.

As a result, mastering engineers have also cut costs and streamlined their operations, which has fortunately come at a time when the equipment costs to undertake digital mastering have lowered, making the prospect of staying afloat more favorable. However, this democratization has removed

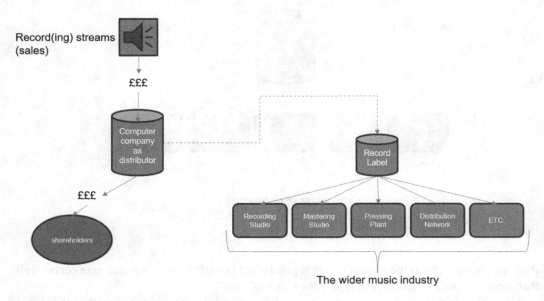

FIGURE 11.4   The contemporary musical economy.

one of the barriers thus reducing any financial security gatekeeping provides. If engineers work at home more now, then what separates a would-be actor from the already established actor?

11.4 Democratization Vs. Gatekeeping

Audio mastering's profile, equipment, and personnel, within the UK music industry, have diversified over time. The field has enjoyed increased popular prominence in recent decades permitting more of those interested in music production to learn about the craft (Nardi, 2014). This has been enabled due to increased discussion, or chatter, whether that be through books, magazines, or the ever-burgeoning wealth of video content online. Prior to this exposure, enabled by the Internet, lay a time when little, if at all, was known about audio mastering outside of those directly involved in the production process or the label. There was very little information in the public domain, or certainly infinitesimally small amounts in comparison to contemporary times. Prior to the new-found exposure, those interested in mastering might not have known it existed at all, let alone generated a deep interest in it.

This, if you like, is a form of knowledge, or information, gatekeeping (Shoemaker et al., 2009, Barzilai-Nahon, 2008). The concept that you cannot ask about something you know nothing about is one phenomenon, the second would be the fact that even if you had conceived there must be a process between the recording studio (mix) and the record on your turntable, you'd be limited to find out the necessary information. Aspiring young people only knew that audio engineering was the known route to a career, and probably likely to be termed a sound engineer rather than the modern all-encompassing term of 'producer' (see Boehm, 2007, 2018). Indeed, beyond this, little was known about how mastering engineers learn their trade.

The introduction of digital audio, digital audio workstations and the large democratization of production technology has, alongside a thirst for content of all forms online, led to the exposure of what mastering is and what is undertaken. It must be pointed out that the term 'undertaken' refers more to an exposure of the equipment used, rather than how to master (Cousins & Hepworth-Sawyer, 2013, Nardi, 2014, Braddock et al,, 2021). In fact, this topic is the basis of my ongoing research and somewhat relates to this discussion.

For those newly interested in audio mastering, practical knowledge is, of course, required, but also knowledge of the equipment. Digital audio technology has provided a larger cross-section of the audio community with access to software emulating the expensive analogue equivalents used by many mastering engineers at a vastly reduced price (in some cases free). The equipment emulations, through Virtual Studio Technology enabled plugins (VST), permit users to undertake mastering actions within their digital audio workstations. Providing their digital-to-analogue conversion (ADC) and their monitoring systems are of a high enough quality, it could be argued they can undertake professional-sounding mastering (Cousins & Hepworth-Sawyer, 2013).

However, gatekeeping remains. Figure 11.5 depicts analogue equipment in the top line indicating a long-standing need to own and maintain high-quality and expensive audio-processing equipment and monitoring just to undertake the act of mastering. Add to this, upon the release of vinyl as a mass-produced method of passing music to consumers, the expensive need for a professional-quality mastering lathe and ancillary equipment. Since 1983, and the introduction of the Compact Disc (CD), there has been a decline in the number of lathes in circulation. The last Neumann lathe, the VMS80, was released in 1980 and there are no industry-wide accepted replacements. Therefore, there are only a finite number of VMS80 lathes in existence alongside earlier VMS70s. This is an added layer of gatekeeping that cannot easily be countered. Figure 11.5 does not reflect

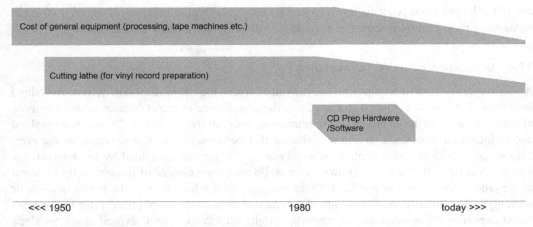

Cost of general equipment (processing, tape machines etc.)

Cutting lathe (for vinyl record preparation)

CD Prep Hardware
/Software

<<< 1950   1980   today >>>

FIGURE 11.5   Historic gatekeeping mechanisms within audio mastering.

the increased demand for the lathes in the past 15 years or so, and the enormous, and rising, price for a VMS80 today. I was intending to buy one in 2010 and was quoted around £30,000 including servicing. Upon recent enquiry, I was quoted something in excess of £100,000 for the same model of lathe. Added to this is the sheer destruction of the machines that were employed to press the vinyl record. As Palm (2017) recounts ". . . during the 1990's most pressing plants were retrofitted to stamp CDs, sold for parts, or simply shuttered" (Palm, 2017).

The release of the CD erupted at a time before hard drive capacities could adequately contain the 650 MB necessary for a whole CD's worth of music. The average capacity of a hard drive in 1983 was in the region of 50 Mb and the cost would have been mortgage-like in scale. A new format had to be considered and engineers. Sony, as co-developers of the CD, repurposed one of their broadcast-level video tape machines by creating a set of convertors (PCM1610 and PCM1630) that changed the analogue audio into digital data that could be recorded onto U-Matic video tapes. Once on the tapes, mastering engineers could compile the data necessary into a flow of audio that would become the CD master which could then be sent by courier to the CD manufacturing plant. Therefore, prices for CD preparation using this system will have been exceptionally expensive in 1983, and by the very cost, provide a gate to the technology. Only labels willing to invest or extremely rich people could afford to have CDs prepared for manufacture.

As technology developed, and hard drive capacities increased above 650 MB, alternatives to U-Matic could be utilized. Skipping a few format steps, a new industry-wide delivery system for CD replication was created by Doug Carson called Disc Description Protocol (DDP). Whatever the format (hard disk, DAT data tape, Exabyte tape, or an FTP server), the DDP protocol would permit the file to be easily error-checked at the plant for integrity and accuracy. Carson freely licensed out the protocol, but many companies used this as an opportunity to make money, and the gatekeeping was maintained. However, over several decades, the DDP became cheaper to engage with in software less-specialist than the mastering DAW packages, hence further democratizing the act of CD mastering.

We have had a period since where the barriers to mastering for CD have all but vanished and this democratization can be considered a positive outcome. However, one indicator that cannot be shown on the image, is there is no longer any barrier for a DAW-based musician releasing directly to a streaming service of choice – effectively removing the perceived need for a mastering engineer.

The only other gatekeeping quality undiscussed thus far is that of knowledge, and with this, experience is included. To be able to master audio requires not only a wealth of equipment at a professional level but also the knowledge to undertake the task. Democratization in this area has been blossoming for several decades now with a range of new books (see Katz, 2003, Cousins & Hepworth-Sawyer, 2013, Hepworth-Sawyer & Hodgson, 2018, Braddock et al., 2021 for a selection). Added to this have been several magazine articles and training specials by the likes of *MusicTech Magazine* dedicated to mastering. Traditional training, within the apprenticeship model of the large studio complex is unattainable to many. Equally, certainly within the UK, there is a void of training directly in audio mastering (Hepworth-Sawyer, 2021).

11.5   Enter Dolby Atmos and Spatial Audio

Until this point in this chapter, I have concentrated on the main delivery presentation of stereo. Music mastering has specifically been dominated in the stereo domain for over fifty years.

The university where I teach undertakes an annual pilgrimage to Studio 2 at Abbey Road with its final-year students. Being a mastering engineer, I've little to do, so have frequently drifted upstairs to speak to those with the same job title. Abbey Road has, of course, been at the forefront of spatial audio installing a Dolby Atmos room several years ago before all the rage and prominence the format has today. Since this time, the mastering team there has developed a small, but perfectly formed, Atmos mastering room. Upon my recent visit to the studio, an engineer, showing me the new room, said that he hoped Atmos mastering would re-instate a gate to the professional industry once more. This comment was the genesis of this chapter.

So, what is the gate that this engineer refers to in real terms? Gatekeeping, as Oleinik stated requires a triad: the established actors (in this case Abbey Road), the would-be actors (engineers currently working in stereo), and a gatekeeper. In the case of Dolby Atmos mastering, the gatekeeping factor is the significant economic investment required to engage in the field. For stereo mastering, only two high-quality loudspeakers are required. For 5.1 (not widely requested for mastering), six speakers are needed. However, for Dolby Atmos, a 7.1.4 configuration is required (or 7.1.2 at the bare minimum). These figures relate to seven speakers in the listening plane, one subwoofer, and four upper speakers to provide the impression of height. Given that stereo mastering monitors are typically in excess of £10,000 a pair, expanding this by another ten mono monitors is a significant investment. Some might argue that this work could be achieved using binaural methods on a pair of headphones. However, discreet drivers within an acoustically treated room will not succumb to the same phase issues binaural exhibits (Figure 11.6). There is yet further expense given the way in which Dolby Atmos requires (in many cases) more than one computer to engage with the process of its creation.

The pressure is not just for engineers to choose to join the race, as they may have done with previous formats (Super Audio CD for example) to find out that the consumer demand is not real. Dolby Atmos specifically, and more latterly Apple Spatial Audio, have been bundled with some of the streaming platforms such as Apple Music, and in such a way that you might not even know you're listening to the spatial edition of your favorite album. Unlike SACD, where the consumer had to make a conscious decision and significant investment to appreciate what the format provided, spatial audio has been covertly unleashed on the listening turning the feature on within an update (www.apple.com/uk/newsroom/2021/05/apple-music-announces-spatial-audio-and-lossless-audio/).

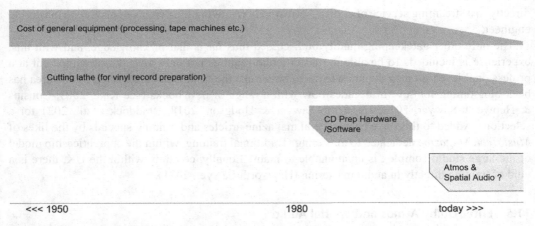

Cost of general equipment (processing, tape machines etc.)

Cutting lathe (for vinyl record preparation)

CD Prep Hardware /Software

Atmos & Spatial Audio ?

<<< 1950   1980   today >>>

FIGURE 11.6   Spatial audio within the historic gatekeeping mechanisms of audio mastering.

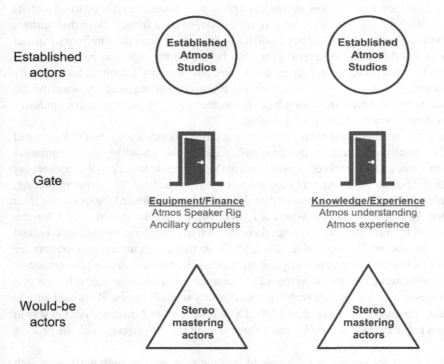

Established actors

Established Atmos Studios

Established Atmos Studios

Gate

Equipment/Finance
Atmos Speaker Rig
Ancillary computers

Knowledge/Experience
Atmos understanding
Atmos experience

Would-be actors

Stereo mastering actors

Stereo mastering actors

FIGURE 11.7   Gatekeeping in spatial audio mastering.

As a result, there is not just pressure from the record labels demanding spatial mixes and masters, but there is, for once, potential pressure from the listening public to engage with spatial audio. Should this pressure become widespread, the engineer from Abbey Road's premonition of a new financial gatekeeper will come true (Figure 11.7). There is some discussion amongst the community that engineers are not being paid much, if at all, for the additional work that spatial audio deliverables cause. I have also heard engineers not receiving the gig for the stereo work, if not also providing the spatial deliverables.

11.6 Gatekeeping Effect On Higher Education

Since the early 1990s, when the second generation of music technologists were emerging and educated at the newly named universities in the UK (post-1992), there have been several educational offerings in the audio field (Boehm, 2007). Despite the void in higher education when it comes to stereo audio mastering education, the core skills of the field have been taught widely (Hepworth-Sawyer, 2021). This has been mainly in the stereo domain, with some universities championing spatial audio over several years, such as the University of Huddersfield with Hyunkook Lee's research, but also other institutions such as Birmingham for example. However, at the start of the widespread consumption of spatial audio by the listening public, there are few educational options to acquire the knowledge to record, mix, let alone master in these new formats. As alluded to before, there's such limited time to gain the necessary listening skills in stereo, let alone in spatial presentations.

In fact, many establishments will be unable to engage in spatial audio en masse, should the format become widespread and permanent. It is not just the prohibitive cost of the equipment alone, but the space required for 11 full-range monitors plus one large subwoofer. Small stereo production rooms, often associated with university facilities, will not be able to cater to this requirement, meaning students may revert to the sub-optimum binaural representations of their work.

11.7 Conclusion

Gatekeeping, in the context covered in this chapter, provides a barrier to engagement that, in turn, results in profits for the companies that can become established actors. Those would-be actors are unable to engage in the art of spatial audio mastering in a way equipment democratization permitted them to in the stereo domain. Whether the gatekeeping effect of spatial audio endures and results in further competition in both the studio and the higher education sector is yet to be realised. What is interesting is the new musical economy, run by data companies, are the instigators of spatial audio upon the masses, not necessarily the recording industry itself. Ventures into surround music presentations before have not succeeded, such as Quad and 5.1, but the difference is the widespread availability of binaural representations without any new equipment. A big shift in the audio mastering industry could be around the corner,

References

Anderson, C., 2007. The long tail: How endless choice is creating unlimited demand. Random House.

Attali, J., 1977. Noise: The political economy of music (Vol. 16). Manchester University Press.

Barzilai-Nahon, K., 2008. Toward a theory of network gatekeeping: A framework for exploring information control. *Journal of the American Society for Information Science and Technology*, 59(9), pp.1493–1512.

Boehm, C., 2007. The discipline that never was: Current developments in music technology in higher education in Britain. *Journal of Music, Technology and Education*, 1(1), pp.7–21.

Boehm, C., Hepworth-Sawyer, R., Hughes, N. and Ziemba, D., 2018. The discipline that 'became': Developments in music technology in British higher education between 2007 and 2018. *Journal of Music, Technology & Education*, 11(3), pp.251–267.

Bourdieu, P., 1984. A social critique of the judgement of taste. Traducido del francés por R. Nice. Londres, Routledge.

Bourdieu, P., 1990. The logic of practice. Stanford University Press.

Braddock, J.P., 2020. Mastering audio analysis: Teaching the art of listening. In Mastering in music (pp.104–118). Focal Press. www.routledge.com/Mastering-in-Music/Braddock-Hepworth-Sawyer-Hodgson-Shelvock-Toulson/p/book/9780367227197

Braddock, J.P., Hepworth-Sawyer, R., Hodgson, J., Shelvock, M. and Toulson, R. eds., 2021. Mastering in Music. Routledge, Taylor & Francis Group.

Cousins, M. and Hepworth-Sawyer, R., 2013. Practical mastering: A guide to mastering in the modern studio. Routledge.

Hepworth-Sawyer, R., 2021. Mastering The Void: A Missing Link in UK Audio Education?. In Audio Engineering Society Conference: 2021 AES International Audio Education Conference. Audio Engineering Society.

Hepworth-Sawyer, R. and Hodgson, J., 2018. Audio Mastering: The Artists: Discussions from Pre-production to Mastering. Routledge.

Hinksman, A., 2020. The creative mastering studio. In Mastering in Music (pp.3–18). Focal Press. www.routledge.com/Mastering-in-Music/Braddock-Hepworth-Sawyer-Hodgson-Shelvock-Toulson/p/book/9780367227197

Hujran, O., Alikaj, A., Durrani, U.K. and Al-Dmour, N., 2020, January. Big data and its effect on the music industry. In Proceedings of the 3rd International Conference on Software Engineering and Information Management (pp.5–9).

Janssen, S. and Verboord, M., 2015. Cultural mediators and gatekeepers. In: James D. Wright (editor-in-chief), *International Encyclopedia of the Social & Behavioral. Sciences*, 2nd edition, Vol 5. Oxford: Elsevier. pp. 440–446. doi:10.1016/B978-0-08-097086-8.10424-6.

Katz, B., 2003. Mastering audio: the art and the science. Focal Press.

Maguire, J.S. and Matthews, J., 2010. Cultural intermediaries and the media. *Sociology Compass*, 4(7), pp.405–416.

MusicWeek, 22 January 2000. *Best sellers of 1999: Singles top 100. Music week*. London, England: United Business Media. p. 27.

Nardi, C. 2014. Gateway of sound: Reassessing the role of audio mastering in the art of record production. *Dancecult: Journal of Electronic Dance Music Culture*, 6, 8–25.

Negus, K., 2011. Producing pop: Culture and conflict in the popular music industry. Edward Arnold.

The Official Chart Company. (2023a) *The best-selling singles of all time on the Official UK Chart*. www.officialcharts.com/chart-news/the-best-selling-singles-of-all-time-on-the-official-uk-chart__21298 [Accessed 11/07/2023].

The Official Chart Company (2023b). *Official Charts Flashback 2003: t.A.T.u. – All The Things She Said*. www.officialcharts.com/chart-news/official-charts-flashback-2003-tatu-all-the-things-she-said__21667/ (accessed 23/08/2023).

Oleinik, A.N., 2015. The invisible hand of power: An economic theory of gate keeping. Routledge.

Page, W., 2021. Tarzan Economics: Eight Principles for Pivoting Through Disruption. Little, Brown.

Palm, M., 2017. Analog backlog: Pressing records during the vinyl revival. *Journal of Popular Music Studies*, 29 (4), e12247.

will.i.am, 2013. will.i.am Interview: Music and Our Digital Future. *Interview with Wall Street Journal* at www.youtube.com/watch?v=0oVcXqB7pnE (Accessed 11/07/2023)

Sanders, A., Phillips, B.J. and Williams, D.E., 2022. Sound sellers: musicians' strategies for marketing to industry gatekeepers. *Arts and the Market*, 12(1), pp.32–51.

Towse, R., 2020. Dealing with digital: the economic organisation of streamed music. *Media, Culture & Society*, 42(7–8), pp.1461–1478.

Wikstrom, P., 2014. The music industry in an age of digital distribution. Change: 19 key essays on how the internet is changing our lives, pp.1–24. www.bbvaopenmind.com/en/articles/the-music-industry-in-an-age-of-digital-distribution/

12

MUSIC MASTERING AND LOUDNESS PRACTICES POST LUFS

Pål Erik Jensen, Tore Teigland and Claus Sohn Andersen

12.1 Introduction

Mastering is the final step in the audio production process. The purpose of mastering is; when needed; to make finishing touches to a final mix seeking to enhance the overall sound, create consistency across tracks, and prepare it for distribution – optimizing playback across all systems and media formats. A central part of the mastering process is deciding the loudness of each track, a process which has become the source of an ongoing debate over several decades, today commonly known as 'the loudness war' – describing a race between audio content producers to be louder than competitors. The loudness war is generally recognized as a result of the 'louder is better' paradigm: "the established assumption that a 'louder' recording will invariably, by comparison, be preferable to most listeners" (Taylor and Martens, 2014).

In 2010, EBU introduced the R 128 recommendation for loudness normalization and maximum levels of audio signals. The aim was to ensure consistent audio levels between different programs and platforms while also countering the loudness war. Now, more than a decade after its introduction, loudness delivery specifications or recommendations and normalization are well adopted on most major platforms for audio distribution. However, when measuring loudness on new charting music today, there seems to be no correlation between the recommendations and the loudness levels of the original mastered files. In fact, music mastering is apparently at least as loud as before. This may imply that loudness normalization has become something for broadcasters and streaming services to deal with, while the music industry keeps on going like nothing has happened. If audio levels are still pushed beyond recommendations while most platforms compensate for the excess levels, a valid question is why? What drives and motivates loudness practices today? Five professional mastering engineers were interviewed with regard to their view on loudness recommendations; industry, client, and consumer expectations; and how these factors inform their mastering practices.

12.1.1 The Loudness War

The term 'loudness war' was used by Robert Orban (Orban, no date) in a 1979 article discussing excessive compression and limiting for FM radio broadcast to boost loudness at the expense of broadcast quality (Orban, 1979). Orban discusses what he refers to as a myth: "that a louder signal, regardless of quality, attracts more listeners" (1979, quoted in Vickers, 2010, p. 3). But

DOI: 10.4324/9781003396710-12

competing about being loudest was not solely a phenomenon of the seventies nor the radio. Devine argues that "loudness has been a source of pleasure, a target of criticism, and an engine of technological change since the earliest days of sound reproduction" (Devine, 2013, p. 156). Thiele writes about 'the jukebox effect' (Thiele, 2005): In 1957 jukeboxes were installed in large quantities in public places where people ate and talked. The customers had no control over the sound level as the system gain was set by the owner and stayed fixed. As a result, recordings with softer levels struggled to compete with high ambient noises from the public areas. So competing recording companies, vying for maximum loudness, applied more and more dynamic compression, producing ever-increasingly louder records (ibid.). Similar to this, "Motown record company adopted a standard called 'Loud and Clear', which used a number of methods to maximize the apparent loudness while maintaining clarity" (Vickers, 2010, p. 3). George Martin, the Beatles' producer, spoke of how baffled he was at how hot American records were and wanted to get the Beatles that sound (Martin and Pearson, 1994). The Beatles in fact lobbied their record label, Parlophone, to get their records pressed on thicker vinyl so they could achieve a bigger bass sound (Southall, 2006). In the documentary *Queen: Days of Our Lives* (O'Casey, 2011), Freddie Mercury and Brian May are seen in the early 70s discussing a desire to get their records as loud as physically possible, hence sending their masters to the US for an American-style loud cut. In the '80s, following the release of the Audio Compact Disc (CD) in 1982, an even more drastic increase in loudness came about (Vickers, 2010; Taylor, 2017), and the term 'the loudness war' eventually became common property. A frequently cited extreme on the loudness war is Metallica's *Death Magnetic* album released in 2008 (Metallica, 2008), often recognized as one of the loudest and most compressed albums in recorded music history (Vickers, 2010, 2011; Devine, 2013; Deruty and Tardieu, 2014; Taylor, 2017). The sound of *Death Magnetic* was considered so offensive and loud that thousands of fans signed an online petition demanding the album be remastered (Smith, 2019).

12.1.2 Loudness Measurements, Recommendations, and Normalization

In 2010, the European Broadcast Union released EBU R 128, a recommendation for loudness normalization and permitted maximum level of audio signals (European Broadcast Union [EBU], 2020). The recommendation states that peak normalization of audio signals has led to considerable loudness differences between programmes and between broadcast channels, which in turn has caused viewer/listener complaints. Furthermore, an international standard for measuring audio programme loudness has been defined in ITU-R BS.1770 (International Telecommunication Union [ITU], 2023), introducing the measures LU [Loudness Unit] and LUFS [Loudness Units, referenced to Full Scale], which will improve the loudness matching of programmes with a wide loudness range. EBU R 128 define -23 LUFS as the recommended loudness target with a maximum peak level of -1dB True Peak for broadcasting (ibid.). Supplementary to R128, in 'AESTD1008.1.21-9 Recommendations for Loudness of Internet Audio Streaming and On-Demand Distribution' (AES, 2021) music loudness recommendations are set to -16LUFS with a loudest track at maximum -14LUFS and a maximum peak level of -1dB TP. Grimm, who analyzed 4.2 million music albums in search of an optimal loudness target for music streaming states in this regard:

> With a loudest track target of -14 LUFS, the majority of albums from the 70's, 80's and 90's can be normalized properly and sound in balance with modern albums. Classical music albums generally have a loudest track at -16 LUFS or lower. These will unfortunately miss normalization

with the proposed target. At the moment, mobile devices do not offer enough gain to enjoy many of the classical music albums on the go.

(Grimm, 2019)

In broadcast, the loudness differences once experienced between different programs and commercials have successfully been stopped following the adaption of EBU R 128 by most broadcasters. Music streaming and On-Demand Distribution services offer the same opportunity, as they too can control playback loudness much in the same manner as broadcasters. Bob Katz makes a point of this in a 2015 article titled; 'Can we stop the loudness war in streaming?'

The rapid replacement of the CD with downloads and streaming brings hope that producers will be able to create the sound of their choice without fear of losing 'the competition' – if the streams, the computer, and portable players institute loudness normalization by default. However, the same loudness war is continuing with downloads and streaming.

(B. Katz, 2015)

In 2016 music streaming and On-Demand Distribution became the major source for listening to music (Recording Industry Association of America [RIAA], 2017). Today, most music streaming and On-Demand Distribution services are conforming to -16 to -14 LUFS as recommended in AESTD1008.1.21-9 (AES, 2021).

12.1.3 Loudness Practices In The Charts Since The '80s

In the book *Mastering Audio: The Art and the Science*, Bob Katz presents the diagram in Figure 12.1, showing the increase in loudness of charting CDs throughout 30 years, from 1980 to 2010 (R. A. Katz, 2015). The graphic illustration shows an 8.8 dB increase in loudness from -17.7 LUFS in 1980 to -8.9 LUFS in 2010. During the 1990s, we start to observe true peaks above 0dBFS, "inevitably accompanied by added distortion, loss of stereo image width, and perceived depth (due to channel correlation when signal density increases)" (B. Katz, 2015, p. 939).

In order to obtain a preliminary understanding of the current status of music mastering in terms of loudness, the 10 most played songs on radio in Norway in 2021 (IFPI Norway, 2021) were

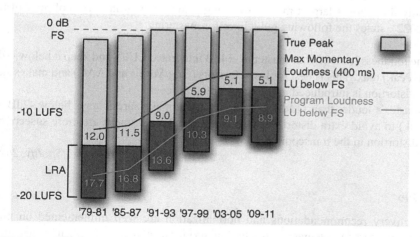

FIGURE 12.1   The increase in loudness of charting CDs throughout 30 years.

TABLE 12.1 Loudness of the 10 most played songs on radio in Norway in 2021

Title	LUFS	dB TP
* Dagny, *Somebody*	-7,1	-0,1
Victor Leksell, *Svag*	-9,5	-0,1
Chris Holsten, *Smilet i ditt eget speil*	-9,9	0,7
* Kygo & Tina Turner, *What's Love Got To Do With It*	-8,4	2,2
The Weeknd, *Blinding Lights*	-8,6	0,7
Ava Max, *Kings & Queen*	-6,1	2,2
* The Weeknd, *Save Your Tears*	-8	0
Ed Sheeran, *Bad Habits*	-6,8	0,1
Olivia Rodrigo, *Drivers License*	-9,8	0
The Weeknd, *In Your Eyes*	-7,5	0,3
AVG	-8,2	0,6

measured for loudness and true peak by the second author. These measurements were not intended to provide a full overview of current practices, but rather to function as an indication and as talking points for the subsequent interviews. Measurement results are provided in Table 12.1.

Each song was played back from Qobuz (*Qobuz*, 2023) using the Qobuz application at the highest audio quality available (44,1kHz/16bit or better) and fed into Pro Tools using the Aux I/O option. Loudness and true peak measurements were performed in Pro Tools using the Nugen Audio MasterCheck Plugin (Nugen, 2022). If the sample rate of the source changed, the 24-bit Pro Tools session would also be changed accordingly, hereby making sure the sample frequency of Qobuz, the Macintosh OS and Pro Tools were always matched when measuring. To validate the measurements, three of the songs (marked with an asterisk in the table) were also downloaded as wave files from Qobuz in their highest resolution. When imported into Pro Tools and measured using MasterCheck, results were identical to the initial measurements performed when streaming from Qobuz.

Comparing the results in Table 12.1 to the charting CDs in the 09'-11' period in Figure 12.1, it is apparent that loudness has not declined over the last decade. If anything, the 2021 measurements are 'worse' showing an increase in integrated loudness from an average of -8.9 LUFS in the 09'-11' period, to -8.2 LUFS in 2021. Furthermore, 6 of the 10 songs in 2021 show true peaks above 0dBFS in our measurements.

Spotify, currently the world's largest music streaming provider, with in excess of 30% of the market (statista, 2022), states the following guidelines for mastering:

- Target the loudness level of your master at -14dB integrated LUFS and keep it below -1dB TP (True Peak) max. This is best for lossy formats (Ogg/Vorbis and AAC) and makes sure no extra distortion is introduced in the transcoding process.
- If your master's louder than -14dB integrated LUFS, make sure it stays below -2dB TP (True Peak) to avoid extra distortion. This is because louder tracks are more susceptible to extra distortion in the transcoding process.

(Spotify, 2023)

12.1.4 Status Quo

While loudness delivery recommendations and normalization are well implemented on most major music streaming and On-Demand Distribution services, there is little to no correlation between the loudness recommendations and the loudness of the mastered music. In fact, music seems to be at least as loud as a decade ago. This may imply that loudness has become something for broadcasters and streaming services to deal with, while the music industry keeps on going like nothing has happened. If audio levels are still pushed beyond recommendations while most platforms compensate for the excess levels, a valid question is why? This chapter seeks to investigate the question at hand by conducting interviews with professional mastering engineers. Findings from these interviews are used to shed light on what drives and motivates loudness practices today.

In the following, the methods used for loudness measurements and the interviews with the professional mastering engineers will be presented before outlining and discussing the results.

12.2   Methods

Five professional mastering engineers based in Norway were interviewed individually. They were informed about the background and purpose of the study and that all personal data would be treated confidentially, not revealing their identity without explicit consent. They were also informed they could withdraw their consent at any time without any given reason. All five engineers completed the interviews and none of them withdrew their consent. Of the five mastering engineers, three have extensive experience with close to, or more than, 20 years of practice, including major acts. For two of the mastering engineers, although having more than 10 years of professional experience in music production and/or audio engineering, music mastering had been their main occupation for the last three years.

The interviews were semi-structured and followed a guide based on questions that were considered of interest for contemporary mastering practices in general as well as on loudness practices in particular, covering the following:

- The subject background and experience
- The mastering process and mastering practices
- Customer expectations and relations
- View on loudness and loudness practices
- Future considerations and expectations

The interviews were conducted in the Norwegian language using the Zoom platform where both audio and video were recorded. Each interview lasted for approximately one hour and was conducted by the first and third authors. The interviews were subsequently transcribed by the first and second authors and coded using NVivo. Coding was done with three objectives in mind; firstly, to extract the essence of each subject's answer to the questions outlined in the interview guide. Secondly, to uncover any commonalities across interviewees' responses, and finally, to uncover any emerging themes not directly related to the interview guide. Methodologically, both the execution and treatment of the interviews draw on narrative inquiry ((Kim, 2016; Riessman, 2008) and grounded theory (Corbin and Strauss, 2015).

12.3   Findings

All five mastering engineers were in favour of loudness normalization asserting it improves the listening experience for consumers. At the same time, none of them consider -14 LUFS as a loudness target for their masters, but as a common target for normalization during playback set by most music streaming and On-Demand Distribution stating the following accounts:

- The current standard, -14 LUFS, may well change in the future. Furthermore, some platforms have their own normalization target.
- Spotify, in addition to allowing normalization to be turned off, also offers subscribers the option to choose from three different normalization levels; 'Quiet' (-19 LUFS), 'Normal' (-14 LUFS) and 'Loud' (-11 LUFS). A song mastered to -14 LUFS will consequently be processed by a "probably crude real time limiter algorithm" (quote from subject) to bring it up to -11 LUFS when normalization level Loud is selected. This will alter the sound and is clearly unwanted. When Spotify simply turns the entire level of a louder master down, it does not alter the sound.
- The current industry practice is to make one final master which is used for all platforms and formats. When needed, due to technical considerations in the format, a different master is produced for vinyl. But the aim is one master for present and future. Informants state that it would be good for business if they were to produce different masters for different streaming services, broadcast, CD, vinyl and so on. But this would become more expensive for the clients and demanding to distribute correctly. The current practice is easy and works fine.
- The current loudness algorithm for measuring LUFS is not perfect. -14 LUFS does not sound equally loud on all music. (And in some cases, like an album, we may seek slightly different loudness throughout as some tracks are supposed to be softer or louder than others). All interviewees agree on the critical importance of making aesthetic judgements.
- Different customers and musical genres call for individual approaches as they have different expectations, needs and traditions.

All mastering engineers support the notion that music is just as loud as a decade ago. However, there seems to be a consensus among the engineers that this is mainly true for some music and genres, especially those aiming for the charts. Different genres have different loudness traditions and contemporary evolving practices according to the engineers. Pop/rock is typically the loudest, and classical music is the softest. Two of the engineers mentioned often not even reaching -14 LUFS integrated when mastering classical music. The three most experienced mastering engineers view the music industry as less concerned with loudness today than before: "The customer rarely asks for loud masters these days, which was common in the '90s and the 20th century" (quote from subject).

None of the mastering engineers interviewed in this study aim at a predefined loudness level in their work. At least two of the mastering engineers do not even regularly use loudness metering as part of their workflow. They rely solely on their familiarity with their listening environment; working at a fixed monitoring level; trusting their musical knowledge and how different music should sound, feel, and be perceived in that environment. The actual loudness level in the final master will inevitably vary based on subjective genre aesthetics and tradition rather than meeting technical specifications, as is the case for all five mastering engineers, as evidenced by the following quotes:

> Telling a passionate artist that their music is not supposed to be louder than -14 LUFS just don't work . . . You can't put such restrains on an artistic expression . . . If they want to be loud, they must be allowed . . . a master at -14 LUFS doesn't sound much urban to me . . . heavy compression and clipping has evolved to a sought-after sound in some genres . . . in which case we had to turn it down just leaving much headroom to conceal it.
>
> *(Individual selected quotations by the interviewed subjects)*

Shifting focus from loudness to peaks; the mastering engineers targets their peak levels to a more predetermined level than is case for the loudness levels, typically between -0,3dB TP and -0,1dB

TP. One of the engineers self-proclaimed that he did not use true peak limiters because he "did not like the sound of them", and as a result, his masters would "occasionally go over". As he always very carefully listens to every finished master as a last precaution before sending it to the client, he could not see any wrongdoing if it sounds as it should. He also stated that "I am obviously not the only one doing so . . . " (quote from subject). Only one of the mastering engineers mentioned the possibility of added distortion in the transcoding process, although all engineers were familiar with the general recommendation of -1dB TP.

None of the mastering engineers think there is still a loudness war going on, at least in the manner we are used to think of it; 'as a competition or a race to be louder than the competitors'. In their view, while some still are concerned with loudness, it is no longer a race. As one mastering engineer put it: "I think using the term loudness war is wrong today, but there might be a 'loudness fear' for some, a fear about not being loud enough, but no real desire to be the loudest" (quote from subject).

12.4 Discussion

There is without doubt a consensus between the mastering engineers in terms of both loudness and peak practices. Although one of the engineers self-proclaimed that his peaks sometimes went over, the majority took some measures to avoid this but nevertheless still pushed levels very close to 0 dBFS. The fact that only one of the engineers mentioned the possibility of added distortion in the transcoding process, but not even he seemed to worry, seems surprising to the authors. On the other hand, the true peak level measurements performed on the 10 most played songs on radio in Norway in 2021, may indicate that this lack of concern is common among mastering engineers. None measured at the recommended -1dB TP. 6 of the 10 songs measured even above 0 dBFS, with true peak levels up to +2,2dB. One question inevitably arises, to which extent are these overs audible? Furthermore, as all the mastering engineers expressed a relatively relaxed approach towards loudness levels, being more concerned with the sound and feel of a production: Why not just lower the peak target to -1 dBFS or more to avoid added distortion due to transcoding and preserve the sound of the master down the distribution chain also in this regard? After all, every engineer recognized music streaming and On-Demand Distribution as the main source for music distribution and listening.

One aspect brought up by some of the mastering engineers was the use of levelling limiters on masters in every part of the production process. To avoid loudness confusion and stay competitive, everyone from songwriters to producers and mixers use limiters to push levels in their mixes. This creates a chain where loudness levels tend to escalate, constantly pushing the loudness of the production. The songwriter(s) push the loudness in the demo to be competitive; the producer pushes the loudness in the rough mix (often used as a reference in the mixing process) to beat the demo; the mixer pushes the loudness in the listening mix to beat the rough. As one of the more experienced mastering engineers working extensively with major acts expressed it: "By the time the production reaches mastering, it is already very loud. When the master is done it might be louder than the final mix, or even slightly softer" (quote from subject). Strongly related to this, all mastering engineers pointed to the fact that mix engineers in recent years have gained increased access to the same tools as the mastering engineer, which in turn has facilitated an increase in signal processing on the master channel in the mixing process. Everyone thought this was just fine as it makes an important part of the sound, but normally sought a mix without a last levelling limiter, leaving more room for the mastering engineer to do their work:

There are many good mixers out there, and their mixes may sound as a finished product. They too make use of a levelling limiter as a last insert in their chain to push loudness on their listening mix used for approval by artist/producer/label. I usually get the listening mix in addition to the final mix (e.g., without the levelling limiter). I naturally want the master to be perceived as at least slightly better than the listening mix. As I always evaluate all the mixes I get, looking for the best starting point; how I might improve on it; I might sometimes end up using the loud listening mix because its already 'that good'. Either way, when working with the best, my masters does not sound much different than the listening mix in the end anyway.

Quote from subject

Some of the mastering engineers mentioned that on the topic of driving or maintaining loudness levels today, the majors must take at least some blame. After all, the loudest music is primarily found in charting music which is a reference for both the established and upcoming. Two of the mastering engineers mentioned that although rarely being asked about a louder master today, these requests are more likely to come from majors. One of the engineers maintained:

There will always be a degree of 'loudness war' present, because as we all know, louder music sounds 'fresher and better'. As an example, when editorial meetings in radio stations decide upon their playlists it might still pay off being perceived loudest and hence freshest.

Quote from subject

Some of the mastering engineers point out that they can observe new trends 'to trick the loudness normalization'. They mention especially louder vocals in the mix, but also adaptations in the arrangement as the use of very soft sections and/or less busy instrumentation as methods which might contribute to the perception of the song being more present/louder/fresher although it measures -14 LUFS integrated. Much in the same way; all mastering engineers constantly evaluate how different productions compares on radio and music streaming, reflecting on what they perceive and analyzing how to 'up their game'. As one of the subjects stated: "Mastering engineers are not high in the food chain. If I don't meet the clients' expectations, it's easy to move the business to someone who will" (quote from subject).

Combining the data from Bob Katz (2015, also see Figure 12.1), the measurements performed by the second author (see Table 12.1), and the statements from the engineers interviewed, the question arises; have the loudness wars come to an end, simply because music, generally speaking, cannot become any louder while also upholding different genre aesthetics?

12.5 Conclusion

All five mastering engineers interviewed in this study are in favour of loudness normalization asserting it improves the listening experience for consumers. At the same time, none of them consider -14 LUFS as a loudness target for their masters, but as a common target for normalization during playback set by most music streaming and On-Demand Distributors.

- All mastering engineers support the notion that music is just as loud as a decade ago. However, there seems to be a consensus among the engineers that this is mainly true for some music and genres, especially those aiming for the charts.
- Different genres have different loudness traditions and contemporary evolving practices according to the engineers. Pop/rock is typically loudest, and classical music typically softest.
- None of the mastering engineers aim at a predefined loudness level in their work. The actual loudness level in the final master will vary based on subjective genre aesthetics and tradition.

- The mastering engineer targets their peak levels to a more predetermined level than is the case for loudness levels, typically between -0,3dB TP and -0,1dB TP. Only one of the mastering engineers mentioned the possibility of added distortion in the transcoding process although all engineers were familiar with the general recommendation of -1dB TP.

The results indicate that contemporary mastering practices are primarily based on subjective genre aesthetics and tradition rather than meeting technical specifications. None of the mastering engineers think there's still a loudness war going on, at least in the manner we are used to think of it: 'as a competition or a race to be louder than the competitors'. In their view, while some still are concerned with loudness, it is no longer a race. As one mastering engineer put it: "I think using the term loudness war is wrong today, but there might be a 'loudness fear' for some, a fear about not being loud enough, but no real desire to be the loudest" (quote from subject).

12.6 Future Works

The audibility of true peak distortion should be investigated further. As some mastering engineers seems to show little concern about this effect it is well worth looking more into. Music mastering is in general viewed as a technical service, but as this study indicates, the engineers' practices are primarily based on subjective genre aesthetics and tradition rather than meeting technical specifications. While one does not exclude the other, it would be interesting examining the work of mastering engineers asking the question; Music mastering, is it the last quality control or the final creative touch?

References

AES (2021) 'Technical document AESTD1008.1.21-9 (supersedes TD1004): Recommendations for Loudness of Internet Audio Streaming and On-Demand Distribution'. Available at: www.aes.org/technical/documentDownloads.cfm?docID=731.

Corbin, J. and Strauss, A. (2015) Basics of Qualitative Research – Techniques and Procedures for Developing Grounded Theory, 4th ed. London: SAGE.

Deruty, E. and Tardieu, D. (2014) 'About Dynamic Processing in Mainstream Music', *J. Audio Eng. Soc*, 62(1/2), pp. 42–55.

Devine, K. (2013) 'Imperfect Sound Forever: Loudness Wars, Listening Formations and the History of Sound Reproduction', *Popular Music*, 32(2), pp. 159–176.

European Broadcast Union [EBU] (2020) 'EBU – Recommendation R 128. Loudness Normalisation and Permitted Maximum Level of Audio Signals'. Available at: https://tech.ebu.ch/docs/r/r128.pdf.

Grimm, E. (2019) 'Analyzing Loudness Aspects of 4.2 Million Musical Albums in Search of an Optimal Loudness Target for Music Streaming', in *Audio Engineering Society Convention 147*. Available at: www.aes.org/e-lib/browse.cfm?elib=20641.

IFPI Norway (2021) Topplista – IFPI. Available at: https://ifpi.no/topplista/ (Accessed: 10 October 2023).

International Telecommunication Union [ITU] (2023) 'ITU-R BS.1770-5: Algorithms to Measure Audio Programme Loudness and True-Peak Audio Level'. Available at: www.itu.int/rec/R-REC-BS.1770-5-202311-I/en

Katz, B. (2015) 'Sound Board: Can We Stop the Loudness War in Streaming?', *J. Audio Eng. Soc*, 63(11), pp. 939–940.

Katz, R. A. (2015) Mastering Audio: The Art and the Science, 3rd ed. Burlington, MA; Abbingdon: Focal Press.

Kim, J.-H. (2016) Understanding Narrative Inquiry: The Crafting and Analysis of Stories as Research. Thousand Oaks, CA: SAGE.

Martin, G. and Pearson, W. (1994) 'With a little help from my friends: The Making of Sgt. Pepper', *(No Title)* [Preprint].

Nugen (2022) 'MasterCheck – Operation Manual'. Available at: https://nugenaudio.com/files/manuals/Mast erCheck%20Manual.pdf (Accessed: 22 July 2023).

O'Casey, M. (2011) Queen: Days of Our Lives. Universal International Music. www.imdb.com/title/ tt1977894/

Orban (no date) Robert Orban. Available at: https://aes.digitellinc.com/speakers/view/767 (Accessed: 5 July 2023).

Orban, R. (1979) 'FM Broadcast Quality', *Stereo Review*, 43(5), pp. 60–63. www.worldradiohistory.com/ Archive-All-Audio/Archive-HiFI-Stereo/70s/HiFi-Stereo-Review-1979-11.pdf

Qobuz (2023) Qobuz. Available at: https://www.qobuz.com/no-en/about (Accessed: 2 August 2023).

Recording Industry Association of America [RIAA] (2017) News and Notes on 2016 RIAA Shipment and Revenue Statistics. Available at: https://www.riaa.com/wp-content/uploads/2017/03/RIAA-2016-Year-End-News-Notes.pdf.

Riessman, C. K. (2008) Narrative Methods for the Human Sciences. SAGE Publications, Inc, Thousand Oaks, CA, U.S.A. and New York for Smith..

Smith, E. (2019) 'Even Heavy-Metal Fans Complain that Today's Music Is Too Loud!!!', in Edited By Theo Cateforis The Rock History Reader. Routledge, pp. 375–378.

Southall, H. (2006) 'Imperfect Sound Forever', *Stylus Magazine*. Available at: www.stylusmagazine.com/ articles/weekly_article/imperfect-sound-forever.htm.

Spotify (2023) 'Loudness Normalization, Spotify'. Available at: https://support.spotify.com/no-nb/artists/arti cle/loudness-normalization// (Accessed: 3 July 2023).

statista (2022) 'Global Music Streaming Subscribers 2022, Statista'. Available at: www.statista.com/statistics/ 653926/music-streaming-service-subscriber-share/ (Accessed: 10 October 2022).

Taylor, R. W. (2017) 'Hyper-Compression in Music Production; Agency, Structure and the Myth That Louder Is Better', *JARP*, 11. www.arpjournal.com/asarpwp/hyper-compression-in-music-production-agency-structure-and-the-myth-that-louder-is-better/

Taylor, R. W. and Martens, W. L. (2014) 'Hyper-Compression in Music Production: Listener Preferences on Dynamic Range Reduction', in *Audio Engineering Society Convention 136*. Available at: www.aes.org/ e-lib/browse.cfm?elib=17169.

Thiele, N. (2005) 'Some Thoughts on the Dynamics of Reproduced Sound', *Journal of the Audio Engineering Society*, 53(1/2), pp. 130–132.

Vickers, E. (2010) 'The Loudness War: Background, Speculation, and Recommendations', in *Audio Engineering Society Convention 129*. Available at: www.aes.org/e-lib/browse.cfm?elib=15598.

Vickers, E. (2011) 'The Loudness War: Do Louder, Hypercompressed Recordings Sell Better?', *Journal of the Audio Engineering Society*, 59(5), pp. 346–351.

Discography

Metallica (2008) Death Magnetic [CD]. Warner Bros.

13

LCR

A Valuable Multichannel Proposition for Modern Music Production?

Juhani Hemmilä and Jason Woolley

13.1 Introduction

Although utilised on successful records, LCR (Left-Centre-Right panning) could be considered a somewhat lost approach within multichannel production techniques. Houghton (2021) recently wrote about LCR technique, but there are limited quality writings otherwise. In a review of ten music production technical and educational books, no reference to LCR was found.

LCR is utilised by practitioners such as Billy Decker and Taylor (2020) and anecdotally is useful in the classroom. Therefore, a possible broader 'reintroduction' of the technique to the educational curriculum might benefit student employability. This project seeks to understand something of whether there are credible reasons, technical or otherwise, that LCR should not be used in production workflows, and therefore not in educational curricula.

Following methods similar to Morell and Lee (2021) and Turner and Pras (2019), the research project discussed in this chapter gauged preferences toward LCR mixes. As the chapter goes on to explain, participants engaged with an online listening survey which included genre-specific examples. Each genre had LCR and full pan versions, and participants were asked to choose their preference. Participants were asked to comment on the reason for their choice.

Placing various sounds in the stereo panorama plays an important role in the process of mixing music. In the early years of record production, technical limitations gave some guidance to where sound 'should' sit in the mix; bass-heavy elements needed to stay in the centre, otherwise the needle of the vinyl player might jump out of its groove (Exarchos and Skinner, 2019). Today, with digital audio, such technical limitations do not apply. Any mix engineer may pan the sounds wherever they see fit, even though most people still place the important parts in the centre (Senior, 2011). Textbooks on music mixing often describe a method of distributing the other sounds in the mix evenly across the stereo panorama in order to create 'a full picture'. Historically, technical limitations of the mix consoles forced engineers at the time to place sounds either in the centre or fully to the left or to the right; there simply was no pan pot. Some of the consoles at that time, for instance the Altec 250SU and the Altec 250T3, were designed to work in LCR mode. The Universal Audio 610 mixing console was also equipped with a Left-Centre-Right switch (Houghton, 2021). Many of the existing standard consoles had to be retrofitted with new electronics for the purpose of creating LCR mixes. Anecdotally, some of the later mixer designs, dating from the late '70s and onwards, do have the capability to do LCR. One example of that is the Soundcraft MH4 console,

DOI: 10.4324/9781003396710-13

introduced in 2004. But the purpose of that design was to distribute audio to three different sets of speakers in a theatre and live music scenario, which is somewhat different from what this research is aimed at.

It is also not uncommon to hear some modern commercial mixes where the mono sounds seem to emerge from only three places: full left, centre and full right (LCR). The places 'in between' are occupied with stereo sounds, for instance drum overheads and synth pads, and stereo reverbs and delays. Some engineers also place mono delays or reverbs in the areas not occupied by mono sound sources. It is important to point out that in this chapter, the concept of LCR does not refer to cinematic LCR stereo sound, where it is common to have a sound system including left, centre and right speakers. The centre speaker in the latter scenario is intended primarily for dialogue. In this research, the focus is on stereo playback of music material. This chapter aims to examine the discrepancy between the lack of information on LCR in textbooks on the subject, and the fact that many modern mix engineers seem to prefer working this way (Houghton, 2021). The aim is to explore if there are perceived technical reasons for not using LCR and favouring 'full panorama' method, and also to investigate whether experienced music engineers have a tendency to prefer one method over the other. The research has implications from a pedagogical standpoint: can increased knowledge of the LCR method give students insights to the mix process and lead to more efficient workflows? Can the demise of LCR mixing or its absence from a great deal of educational literature be explained by aesthetic preferences? Put simply, is LCR not commonly used because practitioners don't like the LCR mixes?

13.2 Background and Literature Review

In preparation for the research, references to LCR mixing technique in literature on music production and music mixing were investigated. Out of the ten books on the subject that the researchers found, none of them had LCR mentioned in the table of contents or index section (Farinella (2006); Alton Everest (2007); Gibson (2007); Bregitzer (2009); Senior (2011); Hepworth-Sawyer and Golding (2011); Moylan (2015); Dowsett (2016); Mynett (2017); Izhaki (2018). All the books do touch upon the topic of panning, but from a traditional viewpoint, i.e. recommending that the full range of pan positions should be utilised. However, Izhaki (2018) approaches the topic stating:

> Multiple mono tracks are those that represent the same take of the same instrument, but do not involve an established stereo miking [. . .] Some of these complementary tracks would be panned to different positions, most likely to the two extremes.
>
> *(Izhaki, 2018, p. 197)*

Izhaki also discusses the concept of masking:

> If each [of two instruments] is panned to a different extreme, they are summed acoustically, which makes masking interaction less harmful. As a result, panning instruments toward the extremes also increases their definition.
>
> *(Izhaki, 2018, p. 199)*

Zager (2006) states the following: "It is helpful to think of the musical canvas [the stereo soundstage] as a clock[, where]. . . the background vocals are spread from 11:45 to 12:15; the snare drum is placed at 11:55 and so forth" (Zager, 2006, p. 120).

A notable exception to the literature, regarding the inclusion of the LCR concept, is the book *Template Mixing and Mastering* by Billy Decker and Simon Taylor (2020). Here, it appears that LCR is a key concept and one of the methods for achieving wide and compelling mixes. Decker relies heavily on templates, as the title of the book indicates. Some insights on the historical reasons for using LCR are given:

> L-C-R [LCR] panning follows a system that was established back in the 1960s as recording studios made the transition from mono to stereo recording. The mixing consoles at the time didn't have panning as we know it today, but instead used a method of sending signals to the left, right, and centre group or master busses.
>
> *(Decker and Taylor, 2020, p. 160)*

Decker goes on to give a rationale for the use of LCR and argues that the technique affords the ability to "throw sounds out as wide as possible". In Decker's view, adding stereo effects pulls the sounds across the stereo sound-field, "blending everything together in a pleasing way" (Decker and Taylor, 2020, p. 85). The British audio technology magazine *Sound on Sound* has two articles on the topic of LCR (Houghton, 2021, 2022). In January 2021, Houghton, in the article 'LCR Panning Pros And Cons', states:

> I. . . would encourage everyone to try it [LCR] at least a couple of times.. . . it doesn't always make mixing easier, but I reckon it does encourage you to make better mixing decisions and often to make them more quickly too.
>
> *(Houghton, 2021)*

In the case of heavy metal, Mynett (2017) states that a contemporary metal production should have lead vocals, bass, kick and snare in the centre and guitars hard-panned left and right. "The 'gaps' within the left-and-centre and right-and-centre are 'filled' by stereo width of the metalwork and tom track, as well by the stereo widening of the reverb and delay processing." Mynett does not, however, relate to this panning strategy as LCR method, even though it appears to be precisely that. There are numerous examples of commercially successful LCR mixes that appear to have been created using LCR panning:

- Perfect Circle (R.E.M, 1983)
- It's Good To Be King (Tom Petty, 1994)
- Vertigo (U2, 2004)
- Don´t Panic (Coldplay, 2001)
- One of Us (Joan Osborne, 1995)
- A Forest (The Cure, 1980)
- Ray of Light (Madonna, 1998)

Whilst a number of recent technical articles have appeared in industry-facing literature such as *Sound on Sound*, including Houghton (2021), there has been limited discourse in academic communities on LCR. Wakefield and Dewey (2015) touched upon the problem of masking in an AES paper. In the latter research, comparison was made of three models for avoiding masking in music mixing. The three methods examined are mirrored equalisation, frequency spectrum sharing and stereo panning. The latter paper's results show that a tool that pans the "target" left and the "masker" right was the most favoured tool by the subjects. Wakefield and Dewey (2015) add a comment from an experienced sound engineer, whose "first choice of technique" for avoiding masking is using panning. In the paper, the researchers do not, however, investigate the concept of LCR mixing. But, the latter results point in an interesting direction.

Broader consideration of academic literature, particularly in relation to participant 'choice' based experiments around multichannel formats, includes Turner and Pras (2019), Malecki et al. (2020), and Morell and Lee (2021). These latter studies considered listener preference in relation to the use of multichannel formats to create immersive experiences.

13.3 Research Methodology

13.3.1 General Research Design

This project utilised mixed methods within the data collection and analysis. Although the sample size of 39 is too small to be significant in terms of the findings, the statistics were found to be useful in helping to identify aspects of the qualitative data, which offered interesting responses to the research question (Creswell and Creswell, 2018).

After some time attempting to develop an online survey that would be fit for the research aims, a decision was made to utilise the JISC platform. The main reason was for ethical clearance from the respective institutions and that the researchers were able to embed media examples that could be played directly from the survey page without redirection. Whilst online surveys are convenient in terms of managing data, they also facilitate a convenient way to broaden the data collection reach beyond local contacts.

The data collection material was primarily a survey with listening examples presented via the JISC online survey platform which contained links to audio examples hosted on the Vimeo platform. Links were seamless and participants were able to engage with the listening examples via the JISC webpages and were not redirected. Each music mix had an LCR version and a 'full pan version'. Participants were asked to listen via headphones in stereo (buds or other head-worn devices) and asked to choose their favourite examples. For each example, once they had made their choice, participants were asked to complete a free text response on why they chose each example.

Participants were allowed to listen to each example as many times as they liked. The survey also collected gender, age, and prior experience information, but all participants remained anonymous to the researchers. This approach aligns with aspects of the methodologies utilised by other authors such as Malecki et al. (2020), Morell and Lee (2021), and also Turner and Pras (2019). Once the data collection was complete the data was analysed through basic collation of the statistical information and through thematic analysis of the qualitative data.

13.3.2 Musical Example Design

In total, 5 pairs of musical examples were created. A folk music example was created and was based on a composition called 'Tora vandrar', by Swedish folk music group Stormsteg (2011). The producer/sound engineer of the album, Hadrian Prett, provided the multitrack recording from which two different mixes were created. The recording was done live in the studio and consists of two violins and an acoustic steel string guitar. The violins were recorded with one microphone each, and the guitar was miked in stereo. A line signal from the guitar was also recorded. Complementing these signals, the producer added two sets of room microphones to capture a natural ambience. Both the research mixes were created from scratch, using mostly the close-miked signals and the line signal.

A jazz example was also created from a multitrack recording of a jazz ensemble performance at the Royal College of Music in Stockholm. The ensemble was led by trombone player Sven Berggren and the other instruments in the ensemble were drums, double bass, vibraphone, piano, saxophone, trumpet, and electric guitar. All instruments were closed-miked, and the double bass also had a line signal recorded.

As an example of the approach to the LCR and 'full' pan stereo mixes, the jazz examples are briefly discussed here. In both mix versions, the drums were in 'true' stereo, with the overhead microphones creating the sound stage and the close mics augmenting the sound captured by the stereo pair. The bass was also in the centre in both versions. In the 'full pan' mix, the three lead instruments trumpet, trombone and saxophone were panned approximately 'quarter to' (trumpet), slightly off centre (trombone) and 'quarter past' twelve. This strategy was intended to create the impression of the players standing in front of the listener as if in a concert scenario. Both the guitar and the vibraphone were stereo recordings and the sounds were distributed evenly full left and full right.

In the case of the LCR mix of the jazz example, the panning was radically different: drums and bass, as mentioned, were the same. But trumpet and trombone were panned full left and full right, respectively, whereas the saxophone was in the centre. The guitar was also panned full left and, finally, the vibraphone occupied the centre, this time in mono. All these five instruments were treated to a touch of reverb. In the case of the horn instruments, stereo reverb. The guitar, however, had a mono plate reverb panned to the opposite direction in order to give a sense of space without cluttering the sound image.

An unpublished pop music genre example was included and this was from a recent production by KMH music production Masters' student Mohamad al Fakir, called 'Gone'. The foundation for the two pop mixes were a large set of stems provided by al Fakir.

The fourth example created was in the style of rock fusion, and is the music of the band Egotrippers (SWE), titled 'Red Light, Green Car' (2022). The first mix version is the original LCR mix found on their album *Space Cruiser Interior Design Revisited* (Egotrippers (SWE) 2022). The second version of the mix is a different 'full range pan-mix'.

Finally, a heavy rock example was created and this was from a performance by Swedish multi-instrumentalist Mattias Ekstam (artist name My Old Habit), who provided the multitrack recording for the creation of the two different mix versions. The heavy rock example is called '. . . Or Words To That Effect' (My Old Habit 2022).

All mixes were created by using Pro Tools Ultimate, various third-party plugins, mainly from Sonnox, and the monitoring used was primarily headphones (Beyerdynamic DT880 Pro) and occasionally also on Genelec 8030 loudspeakers. The mastering was done using plugins on the output in Pro Tools in conjunction with Cloudbounce, an AI-based automated mastering application.

In the process of mixing, a decision was made not to alter parameters such as channel levels, EQ, and compression between the different mix versions. In this project, the decision was made not to alter things other than pan positions as this might run the risk of creating research noise. A similar strategy appears to have been chosen in Morell and Lee's (2021) research on binaural mixing of popular music, in which different versions of three types of music were created. The basis for Morell and Lee's mixes were multitracks that were sent to buses containing plugins for creating different types of binaural effects. But besides panning and binaural treatment, no changes appear to have been made using compression and EQ on individual tracks.

A final point on the LCR versions would be that in addition to panning the dry tracks to either full left or full right, or centre, where mono tracks were treated to mono reverbs and/or delays, these auxiliary tracks were placed in other positions to the original 'dry' tracks the auxiliaries were derived from. Continuous checks in mono were made to ensure that the balance would not change drastically between mono and stereo playback. For links to the music examples, please contact the authors.

13.4   Discussion and Analysis Of Results

13.4.1   Quantitative Data

It appears that the preferences for or against LCR panning depends on many factors. Primarily, the genre in question seems to dictate if a LCR method is appropriate. But having said that, some of the results appear surprising. The pie charts in Figure 13.1 show the preference for LCR panning showed in dark grey colour, whereas the light grey colour shows the preference for traditional 'full' panning.

What came as a surprise in the survey, was that a majority, 59%, of the participants preferred the LCR mix of the folk music example (see Figure 13.1). This type of music is in most cases mixed in the 'traditional' way, i.e. panning the instruments so that they represent a natural soundstage, as if the musicians are in front of a sound stage. The researchers were even somewhat reluctant to use this LCR mix in the survey, because at first, it seemed to be too wide and somewhat unnatural. But the free text responses show that this was exactly what some of the participants enjoyed with the mix. Here are a few comments from participants who preferred the folk music LCR example:

> Better separation between instruments and nicer proximity which improves dynamics and harmonics. Also wider stereo field with improved clarity.
> I like the first version [LCR mix], it sounds wider and you can hear the instruments more clearly separated.

Comments from some of the participants who preferred the full pan version highlight an interesting aspect:

> When listening in headphones wider panning makes it the instruments disjointed. When listening in speakers that's not really a problem.

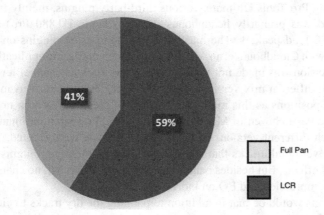

FIGURE 13.1   Folk music.

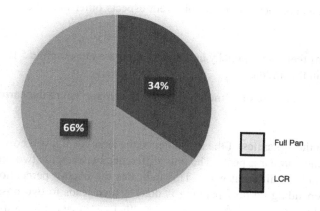

FIGURE 13.2 Jazz.

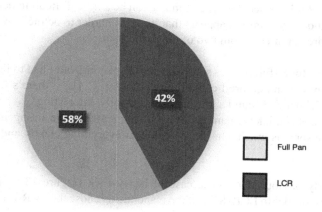

FIGURE 13.3 Pop.

In the case of the jazz music example (Figure 13.2), participants tend to prefer the full pan, or 'normal', type of panning scheme. 34% of the responses were in favour of the LCR method. It appears that many in favour of full pan do appreciate the upside of the LCR mix, but still prefer the aesthetic the full pan version offers. Interestingly there was no consensus on the use of the term 'natural' and this was applied to both versions.

Example 1 [LCR] has a more natural stereo spread.

The second [full pan] one sounds more natural.

The pop example (Figure 13.3) gave somewhat similar results to the jazz example. 58% preferred the full pan version, 42% chose the LCR version. In the text comments opinions differ quite a lot, and some of the participants focus on many different aspects other than the panning method being used.

> The mix on example 6 [full pan] feels more punchy and driven compared to example 5 [LCR]. Example 5 was very squashed in the middle.

> Example 5 [LCR] felt wider and it is more of a 'massive' sound in comparison to the narrower example 6 [full pan].

Some comments were received on the dynamics of the instruments, the lead vocal level, how well the mix is "glued" together and other similar aspects. The only difference between the two mixes, however, is that the panning is made in different ways. This indicates the overall perception of a mix can change dramatically depending on how panning is done, which leads to two possible process considering conclusions: 1) do the panning first and adjust other parameters afterwards or 2) fine-tune the panning to accommodate for changes already made to other parameters like EQ and compression.

When it comes to the rock fusion example (Figure 13.4), opinions are very even: 51% preferred the full pan mix, 49% chose the LCR mix. Since this type of music can be a little difficult to define genre-wise, thanks to its somewhat quirky nature, it appears that it is also harder to define how the panning 'should' be done. Below are comments from two of the participants:

> I noticed I had a slight preference for 8 [full pan] as soon as it started playing, but I had to listen to both examples a couple of times again before I could figure out why . . . []. . . there's some magic in this juxtaposition [LCR] which I found a bit surprising actually. Interesting.

> I prefer Ex. 7 mix [LCR] because I think the panning of the instruments suit the arrangement of this song. Ex 8 didn't lift the song as much. On Ex. 7 I can hear everything clearly and the mix still has cohesion in it.

In the research, it became fairly obvious that in the case of heavy rock (Figure 13.5), it is expected to have a panning strategy similar to LCR. 82% of the respondents chose the LCR version

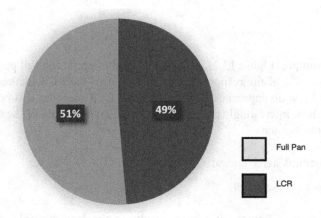

FIGURE 13.4 Rock fusion.

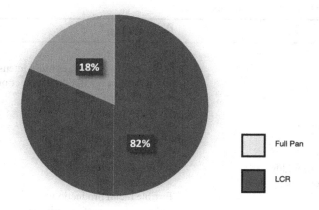

FIGURE 13.5 Heavy rock.

over the full pan version. It seems that listeners like the music when it is as wide and exciting as possible. This insight would most probably lead to certain decisions in the production process. It is, for instance, more or less mandatory to doubletrack the rhythm guitars in this heavier kind of genre. This also leads to the observation that it is no wonder that many heavy metal bands feature a lineup consisting of a least two guitarists. As a side note, in later lineups of British heavy metal band Iron Maiden, there are in fact three guitar players. Interestingly, on their latest album *Senjutsu*, these three guitarists are also panned LCR, respectively (Iron Maiden, 2021).

It is worth noting that the instructions preceding the online test stated that listening should be undertaken using headphones. But it appears some participants did in fact also listen to the music in speakers, which does lead to a somewhat different experience.

One participant comment indicates that there are cons with both mixes:

I don't like any of them but prefer the second one. The first one is too wide and doesn't sound natural, it has lost the bodies of the instruments. The second one sounds more natural but is a bit boxy and closed in. My opinion would probably change if I knew more about the recording and the goal with the audio.

13.4.2 Qualitative Data

To develop a response to the aims of the project, the free text responses collected were subject to thematic analysis. This was a reflexive approach and the coding of responses into themes was undertaken with an awareness of the project standpoint and the influence this inevitably has on the establishment of themes (Ayre and McCaffery, 2022). The authors felt that establishing two main themes of 'technical' and 'emotive' would help understand the two main questions on whether LCR has been a less used technique, and therefore discussed less in educational literature because of either technical or 'emotive' reasons. Technical refers to the use of language commonly used in music production-related discourse to refer to specific technical reasons for why a technique or process is not working. Emotive refers to those responses that are weighted toward a more subjective personal listening preference, but where the participant's response appeared to place

TABLE 13.1 Thematic analysis for the heavy rock examples

Example	Technical theme	Emotive theme
Heavy rock	Controlled	Sounds like crap
	Present	Music painting becomes more intense
	Separated	mix is lame compared to the other one.
	Muddiness	Did not like
	more clearer	More perspective
	more panning is ok	sounded better
	The balance	I feel like is lacking
	a bit more "mono"	more distinctive,
	space and a sense of width	Sense of envelopment
	clear centre to focus on.	sounds better
	wider sound	Terrible metal production
	wide and fat	Did not like any of the mixes really
	width and beefiness	boring
	Feels punchier	
	tighter for me	
	slightly crisper sound	
	panning . . . ok . . . for this kind of instrumentation	
	clearer	
	Better separation	
	more "mono"	
	Wide but with a clear centre to focus on	
	more widely panned.	
	better balance	
	better balance	

lesser importance on providing a technical reason for the choice made. It should be noted that whilst some responses were either wholly technical or wholly emotive, there were also some responses that were a combination of the two.

For the purpose of demonstration, Table 13.1 shows the thematic analysis of free text comments received regarding the heavy rock examples. Table 13.1 shows the key words and phrases used to categorise into either 'technical' or 'emotive' themes. This process involved placing the response into the theme that most characterised the overall response. With regard to the thematic analysis applied, Table 13.1 shows that the majority of rationales given for choosing a particular example could be considered 'technical' in theme.

Table 13.2 above shows that in the instance of the jazz examples, the majority of rationales given for choosing a particular example could be considered 'emotive' in theme. The balance between 'technical' and 'emotive' rationales did vary a little between the other genres presented. Folk and pop rationales were generally evenly split between the two themes, but there were more 'technical' rationales than 'emotive for the rock fusion examples.

In addition, a third separate qualitative theme was established for where negative rationales for the choices were made. This was to try to further understand whether preferences were positively based, or whether sometimes both examples were disliked, and the least worst was chosen. Many rationales given for choices were articulated using 'positive' comments on the chosen version. This was as opposed to other instances where rationale for choices was articulated through negative

TABLE 13.2 Thematic analysis for the jazz examples

Example	Technical theme	Emotive theme
Jazz	more air and room	more natural
	more clearly and present	very unnatural
	better blend for the instruments playing in unison	some weird things going on
	Better balance between the instruments and not so wide stereo	Better organic sound, more "Jazz"
	Some parts seem clear[er] in Example 4	A little more "natural" and more "Jazz"
	Example 4 because it's less brass going on hard pan Left and right	Its more interesting
	more glued together.	more annoying
	More coherent, more focused.	natural stereo image for this type of music.
	clearer, more balanced.	more enyojable for me [sic]
	more narrow	more "together"
	too wide	felt more for ex 4
	better tone	felt more like it would sound hearing the live
	not so spread out (heavily panned))	more like a traditional Jazzmix
		felt better to listen to
		Something felt warmer
		I didn't like
		To me, both example 3 and 4 sound pretty good.
		better positioning of the instruments for my liking.
		feels unnatural for the style of music
		sounds more interesting to my ear
		just felt better to listen to
		No 4 is nicer
		the panning drives you crazy.

comments on the alternative version (e.g. '[Example 2 (folk)]. . . feels less messy'). The latter analysis revealed twelve examples of 'negative' rationales for choices made.

The thematic analysis of the data collected also revealed specific terms that relate to the work done on 'liveness' by Auslander (2008) and also in relation to jazz, by Woolley (2019), regarding ideological notions of authenticity. It was noted that the words 'natural' and 'unnatural' usually appeared in the free text responses to the jazz and folk examples. This is with the exception of the rock fusion example, which did have two comments using the term 'natural'. See Table 13.3.

When considering 'natural' and 'unnatural' comments received, for the jazz examples there was no clear consensus on which example, LCR or full pan, sounded 'natural'.

To summarise the qualitative findings, when considering the themes established, the ratio of emotive rationales, compared to technical rationales does vary depending on genre. The limited number of thematically coded 'negative' responses contributes to an understanding that choices were not coloured by a general dislike for both mixes. The themes as presented here do not indicate a significant and dominant rationale, technical or emotive, for either choosing or rejecting a particular panning method over another.

TABLE 13.3 The use of the terms natural and unnatural

Genre	Natural	Unnatural
Folk	5	0
Jazz	5	1
Rock fusion	2	0

13.5 Conclusions

In conclusion, the data collected and analysed for this project appears to indicate that LCR can be a viable and valuable method for creating acceptable modern musical productions. The discussion has highlighted that there was no consensus with regard to any of the examples where LCR was unfavoured as a choice, and generally speaking the limited statistics point to an even split in preferences between 'full pan' and 'LCR'. It was noted that the heavy rock LCR example was a more popular choice and this was expected due to the predominance of LCR in earlier and some current production of the genre. A few surprises were also noted in relation to LCR being preferred by some participants in the case of genres normally more associated with a 'live' aesthetic and presentation, such as folk and jazz.

The qualitative data and analysis has usefully revealed some reasons for the specific choices made and has provided some insights with regard to whether choices were 'technically' driven or were 'emotive' in character as outlined in the discussion. The latter thematic analysis was also useful in addressing aspects of our research questions and specifically on whether LCR has become a lesser-discussed technique within educational and industry literature because it is somewhat practically and/or creatively redundant in the context of modern production. An explanation of why the lack of LCR discussion in the educational literature highlighted earlier is beyond this chapter. However, a speculative suggestion with regard to the lack of visibility of LCR in literature might be that commercially available hardware and software console designs are usually 'full' pan design, and therefore LCR is not intuitively laid out in the systems most modern music producers are using to produce. LCR switching systems on consoles appear predominantly a 'vintage' technology. Most modern commercial hardware console designs incorporate 'full' pan pot designs or an LCR format designed for Centre speaker applications for screen-based audio post/ or live production workflows. Furthermore most, if not all of the popular DAW software mixer designs incorporate digital interfaces with representations of 'full' pan pot designs. These latter points and the emergence and promotion of new technologies such as the current immersive formats have perhaps overshadowed LCR as an approach.

The research conclusions support our experiences of the usefulness of LCR as an approach to music production and as a pedagogical tool. Our experience has been that LCR offers a pathway to speedier workflows and creative satisfaction, and in the classroom, this has the potential to improve the student experience. This research demonstrates support for the LCR aesthetic and therefore supports our ongoing pedagogical plans.

Alongside the size of the sample, a final note on limitations to mention is the language used in the free text responses. The majority of the text replies were written by participants for whom English is at least a second language. A good portion of responses were also translated from Swedish to English. The conclusions are therefore presented with all the outlined limitations.

13.6 Future Work

With regard to future work, further exploration of the rich data set collected to explore aspects such as the impact of age on mix preference is planned.

This initial project is part of a larger series of research activities underway that explore music production workflow and that we hope will make a positive contribution to industry and educational practice.

References

Alton Everest F. (2007). *Critical Listening Skills*. Boston, MA: Thomson Course Technology.

Auslander, P. (2008). *Liveness: Performance in a Mediatized Culture*. 2nd Edition. Oxfordshire: Routledge.

Ayre, J. and McCaffery, K. J. (2022). 'Research Note: Thematic analysis in Qualitative Research', *Journal of Physiotherapy*, 68, pp. 76–79. DOI:10.1016/j.jphys.2021.11.002 [Accessed August 2023].

Bregitzer, L. (2009). *Secrets of Recording*. Oxford: Focal Press/Elsevier.

Creswell, J. W. and Creswell, J. (2018). *Research Design: Qualitative, Quantitative, and Mixed Method Approaches*. 5th edition. Los Angeles: SAGE.

Decker, B. and Taylor, S. (2020). *Template Mixing and Mastering*. Ramsbery, UK: The Crowood Press Ltd.

Dowsett, P. (2016). *Audio Production Tips*. New York, NY: Focal Press/Taylor and Francis.

Exarchos, M. and Skinner, G. (2019). 'Bass | The Wider Frontier: Low-end Stereo Placement for Headphone Listening', In Gullö, J.O., Rambarran, S., & Isakoff, K., (Eds.), Proceedings of the 12th Art of Record Production Conference Mono: Stereo: Multi (pp. 87-103). Stockholm: Royal College of Music (KMH) & Art of Record Production. Available at www.arpjournal.com/asarpwp/bass-the-wider-frontier-low-end-stereo-placement-for-headphone-listening/ [Accessed August 2023].

Farinella, D. J. 2006. *Producing Hit Records*. New York, NY: Schirmer Trade Books.

Gibson, W. A. (2007). *Mixing and Mastering*. New York, NY: Hal Leonard Books.

Hepworth-Sawyer, R. and Golding, C. (2011). *What is Music Production?* Burlington, MA: Elsevier.

Houghton, M. (2021). 'LCR Panning Pros And Cons, A Mixing State Of Mind', *Sound on Sound Magazine*. Available at www.soundonsound.com/techniques/lcr-panning-pros-and-cons [Accessed August 2023].

Houghton, M. (2022). 'How to Achieve Better Separation in Your Mixes', *Sound on Sound Magazine*. Available at www.soundonsound.com/techniques/how-achieve-better-separation-your-mixes [Accessed August 2023].

Izhaki, R. (2018). *Mixing Audio*. New York, NY: Routledge.

Malecki, P., Piotrowska, M., Sochaczewska, K., and Piotrowski, S. (2020). 'Electronic Music Production in Ambisonics-Case Study', *Journal of the Audio Engineering Society*, 68(1/2), pp. 87–94 (January/February). DOI:10.17743/jaes.2019.0048 [Accessed August 2023].

Morell, P. A. and Lee, H. (2021). 'Binaural Mixing of Popular Music', *Audio Engineering Society Convention e-Brief 665*. Available at www.aes.org/e-lib/browse.cfm?elib=21528 [Accessed August 2023].

Moylan, W. (2015). *Understanding and Crafting the Mix*. New York, NY: Taylor and Francis.

Mynett, M. (2017). *Metal Music Manual*. New York, NY: Taylor and Francis.

Senior, M. (2011). *Mixing Secrets for the Small Studio*. Burlington, MA: Elsevier.

Wakefield, J. P. and Dewey, C. (2015). 'An Investigation into the Efficacy of Methods Commonly Employed by Mix Engineers to Reduce Frequency Masking in the Mixing of Multitrack Musical Recordings', *Audio Engineering Society Convention e-Brief 9341*. Available at www.aes.org/e-lib/browse.cfm?elib=17765 [Accessed August 2023].

Woolley, J. (2019). 'The cultural politics of using technology to support the aesthetic in jazz record production', In Gullö, J.O., Rambarran, S., & Isakoff, K., (Eds.), Proceedings of the 12th Art of Record Production Conference Mono: Stereo: Multi (pp. 329-346). Stockholm: Royal College of Music (KMH) & Art of Record Production. Available at www.arpjournal.com/asarpwp/the-cultural-politics-of-using-technology-to-support-the-aesthetic-in-jazz-record-production/ [Accessed: August 2023].

Zager, M. (2006). *Music Production: A Manual for Producers, Composers, Arrangers, and Students*. Lanham, MD: Scarecrow Press.

Discography

Coldplay, (2001), [CD] *Don't Panic*, Parlophone.

Egotrippers (SWE), (2022), 'Red Light, Green Car', *Spaceship Interior Design Revisited*. Available at: Spotify (Accessed 3 February, 2023).

Iron Maiden, (2021), [CD] *Senjutsu*, Parlophone Records.
Joan Osborne, (1995), 'One of Us'. Available at: Spotify (Accessed 25 May, 2023).
Madonna, (1998), [CD] *Ray of Light*, Warner Bros. Records.
My Old Habit, (2022), '. . . Or Words To That Effect'. Available at: Spotify (Accessed 22 January, 2023).
R.E.M, (1983), 'Perfect Circle'. Available at: Spotify (Accessed 5 June, 2023).
Stormsteg, (2011), 'Tora vandrar'. Available at: Apple Music (Accessed 14 March, 2023).
The Cure, (1980), 'A Forest'. Available at: Spotify (Accessed 25 May, 2023)
Tom Petty, (1994), 'It's Good To Be King'. Available at: Spotify (Accessed 7 June, 2023).
U2, (2004), 'Vertigo'. Available at: Spotify (Accessed 25 May, 2023).

14

RETHINKING IMMERSIVE AUDIO

Adam Parkinson and Justin Randell

14.1 Introduction

Immersion has become a buzzword within arts, technology and academia. Experiences in concert halls, clubs, galleries and museums are often presented and marketed as being immersive, with numerous public and private bodies providing funding for immersive research and technologies. It is a frontier of innovation in music, where emerging technologies facilitate novel experiences. However, immersion itself is defined in different ways by and even within different disciplines and remains a slippery term, with some suggesting it has become a vague, diluted and all-inclusive concept which needs re-examination to remain useful (Agrawal et al., 2020, McMahan, 2003, Thon, 2008).

Within music and sound, "immersive audio" has become synonymous with multichannel audio, namely sound coming out of three or more speakers that surround the listener(s), or sound rendered binaurally through headphones to give the same impression. Immersive audio formats stem from a range of disciplines, including broadcast, cinema and music, using spatial audio technologies such as d&b Soundscape, NHK 22.2, Dolby Atmos, Ambisonic, Sony 360 Reality Audio and 4D Sound. As is argued in this chapter, whilst playback technologies may be a significant component of immersive auditory experiences, immersive audio should not be *reduced* to these technologies.

It is proposed here that our understanding of immersive auditory experiences can be enriched by examining how different disciplines have defined and worked with the concept of immersion. Drawing primarily on theatre, heritage and gaming, some key facets of immersion that researchers and practitioners working in these disciplines have proposed are considered. A series of insights pertaining to immersive experiences drawn from these disciplines are then discussed, with the hope that reflecting on sound practices by way of these cross-disciplinary insights can provide novel ways of approaching immersive sonic experiences. Immersive audio is then considered in the broader context of the UK's so-called "immersive industries", speculating on how this context informs what is thought of as being immersive and could lead to reductive, technology-centric conceptualisations of immersive audio.

The research underpinning this chapter was prompted by the authors' creative practice in the form of *Lorenz Factor*, an immersive electronic music performance which uses audience interaction along with spatialised sound. Developing this piece provided a catalyst for reflecting on how factors such as staging and interactivity contribute to immersion in a performance context.

DOI: 10.4324/9781003396710-14

Amongst the premises that underpin this chapter is the idea that technology alone does not guarantee audience immersion and that any musical experience, regardless of the technologies involved, has the potential to be immersive. This chapter therefore is an attempt to extend technology-focused definitions of immersive audio and explore the range of factors, identified in other disciplines, that can make an experience immersive. Rethinking immersion in this manner may provide inspiration for creative practitioners working with sound and expand the conversations around immersive audio.

This chapter does not provide a recipe for immersive audio experiences, nor is any attempt made to rigidly define immersion. Instead, immersion is embraced as a multiplicity, and the different conceptualisations of immersion challenge fixed or reductive understandings of the term.

A full synthesis of the literature on immersion across disciplines is far beyond the scope of this chapter, and drawing on a range of disciplines outside our own inevitably risks over simplifications, misconstruing ideas and diluting concepts as they are taken out of context. Despite these limitations, it is hoped, however, that this bricolage does no disservice to the scholars that are drawn upon, and that the insights presented can provide germs of inspiration for anyone working with immersive sound.

14.2 Lorenz Factor

14.2.1 Background

The catalyst for this chapter was *Lorenz Factor*, a performance by the authors originally presented at the *Everyday is Spatial* conference at the University of Gloucester in June 2022, and later performed at the *Innovation in Music* conference at Edinburgh Napier University in July 2023. *Lorenz Factor* is a semi-improvised electronic music performance with spatial sound for an assemblage of laptop, modular synth, Max patches, webcam and hardware. The performance also has a dimension of audience interactivity, through a system which tracks audience movements and uses this data to control timbral and spatial elements of the sound. The process of preparing, performing and reflecting on this piece brought us to consider more broadly the idea of "immersion".

The development and performance of this piece across different scenarios led us to start thinking about what it was that made musical experiences immersive. Was it just the spatialisation and multi-speaker playback system that made this, and would the piece fail to be immersive if it was rendered in stereo? To what extent did the participatory elements make the piece more or less immersive? Thinking through these questions led us to review immersion across different disciplines, and begin to map out the range of factors that may contribute to an immersive musical performance.

14.3 Immersion In Sound and Beyond

Immersive experiences are not confined to music, and are found in gaming, theatre, heritage, VR, art galleries, theme parks and more. Examples include Punchdrunk's immersive theatre pieces such as *The Masque of the Red Death* and immersive heritage experiences such as *London Mithraeum* or *The Gunpowder Plot*; video games that are widely acclaimed as being immersive include *The Elder Scrolls V: Skyrim*, *Red Dead Redemption 2* and *Fallout 4*. In the past few years there have been numerous immersive art exhibitions including *Van Gogh: the Immersive Experience* (multiple sites) and *David Hockney: Bigger & Closer (not smaller & further away)* (The Lightroom, London).

Within these disciplines immersion is defined differently and at times without agreement, as has been pointed out by Agrawal et al. (2020), Lee (2020), Biggin (2017), Brown and Cairns (2004),

Rebelo (2021) and others. Nonetheless, scholars and practitioners within these disciplines have attempted to find out what it is that makes a theatrical performance, game or museum visit an immersive experience for audiences, and these different perspectives on immersion can enrich an audio-centric perspective. The following section explores the ways in which immersive experiences are (broadly) defined in different domains.

14.3.1 Gaming

A significant amount of research into immersive experiences comes from gaming. Whilst newer games that use technologies such as VR and AR may be the most obviously immersive, many video games create immersive experiences for their players. The nature of this immersion is contested with different models being presented. Brown and Cairns (2004) used Grounded Theory and interviews with gamers to present a model for immersion which sees it as a continuum of engagement and involvement with a game, with total immersion resulting in equating to feeling completely present in the game. Ermi & Mäyrä (2005) noted that there are different types of immersion, identified through their SCI (Sensory, Challenge and Imagination) model. Other categories of gaming immersion proposed include spatial, ludic and social (Thon, 2008). Attempts have also been made to objectively measure gaming immersion through tracking participants' eye movements and ability to switch tasks, the results of which tend to support subjective measurements of immersion (Jennett et al., 2008).

14.3.2 Theatre

The history of immersive theatre dates back to at least the early twentieth century but is most recently associated with the works of Punchdrunk, formed in 2000 (Biggin, 2017). Contemporary immersive theatre regularly involves interactive elements, and the audience tend to be active participants rather than passive spectators. Fast Familiar's *The Evidence Chamber* places the audience in the role of a jury who are able to shape the outcome of a trial (and thus the direction of the play). Immersive theatre often involves a different approach to space, sometimes using unconventional venues, such as Punchdrunk's *The Masque of the Red Death* utilising the Battersea Arts Centre. In her book on Punchdrunk and immersion, Rose Biggin (2017, p. 2) neatly summarises the key elements in immersive theatre as such:

> The ability of an audience to wander with apparent freedom through a spatially innovative environment, usually scenographically rich and multisensory; a non-chronological and/or impressionistic approach to narrative; and interactive elements or characters, often with an emphasis on empowerment, choice or freedom for the spectator.

14.3.3 Heritage

There is a tradition of immersive heritage experiences in the UK dating back to at least the 1980s, though this has seen rapid expansion as part of the "experience economy" in the past decade or so (Sterling, 2020). Kidd (2018) suggests that characteristics of immersive heritage experiences include being "story-led, audience and participation centred, multimodal, multisensory and attuned to its environment". This is bound up with a number of recent developments in the heritage sector, such as a "narrative turn", an "affective turn" and a "ludic turn", which have seen efforts to include a plurality of voices in museological experiences as well as explore gamified, interactive elements. According to Kidd and McEvoy (2019): "Arts and heritage institutions hope immersive experiences will lead to increased visibility and a culture of innovation, new audiences, more meaningful participation, better engagement, and additional revenue." Put more cynically, "Doing 'something immersive' is increasingly seen as a way of maintaining relevance and securing visibility in a crowded and complex content landscape" (Gröppel-Wegener and Kidd, 2019).

14.3.4 Virtual Reality

Immersion in VR is commonly equated with a feeling of presence, with the two terms often being conflated or used interchangeably (Wilkinson et al., 2021, p. 202). Presence in this context is generally understood as the "feeling of being there" (McMahan, 2003, p. 68). This has origins in the historical connections between VR and telepresence, and the efforts to make users of VR feel physically present in a place different to the one in which they physically reside (Lombard and Ditton, 1997).

14.4 Insights

Let us now discuss what scholars and practitioners within these aforementioned disciplines consider to be some of the key factors in creating immersive experiences, and reflect on how these insights could inform audio practices.

14.4.1 Insight 1: Decouple Form From Experience

Biggin (2017, p. 4) proposes decoupling *form* from *experience*, noting: "The word immersive, therefore, might describe the shape or genre of a production, or the emotional quality of experiencing it." This distinction between immersive meaning psychological states of individuals and immersive meaning objective properties of technologies is also discussed by Agrawal et al. (2020). With "immersive audio" often used to refer only to immersive forms, this decoupling of form from experience can be a very useful critical tool. Immersive forms can include binaural mixes, ambisonic speaker arrays or Dolby Atmos cinemas, and immersive audience experiences would depend on the subjective experience of audience members.

On the connection between form and experience, Biggin (2017) notes that whilst barriers to immersion can be removed, immersive experiences cannot be guaranteed. An immersive playback system may help the audience have an immersive experience, but it won't guarantee it, nor is it even essential for an immersive experience. This decoupling is a reminder to think of immersion not just from the perspective of the form or performer, but from that of the audience. Perfection of the immersive form only goes so far in enabling immersive experiences. Other factors such as the audience's predisposition, sociological factors and the physical environment can have a significant impact on whether the experience is actually conducive to audience immersion. Cultural factors can also have a significant impact on what an audience considers "immersive" as the research of Kim (2021) identifies, showing that North American and Japanese test subjects found different speaker configurations to be more "immersive", with North Americans favouring a frontal, narrow image and Japanese preferring a wide, enveloping image. It is important to think holistically about immersive forms, acknowledging how the staging of a piece, the lighting, the acoustics and general feel of a space all impact immersion.

14.4.2   Insight 2: Think (Critically) About Participation

Across disciplines, a link is often drawn between immersion and interaction or participation. Whilst participation and interaction are distinct phenomena, they are linked by the idea of audience members having a level of active involvement with the piece.

An immersive heritage experience, as described by Kidd (2018), often foregrounds audience participation. Writing on video games, Collins (2013, p. 141) argues that immersion emerges from interaction with a game and that "the act of play, including content creation, leads to the immersive experience". Van Elferen's (2016) ALI (Affect, Literacy, Interactivity) model for analysing immersion in game music also presents interaction as playing a significant role in player immersion. Writing on VR, Bucher (2017) describes how immersion is "less about telling the viewer a story and more about letting the viewer discover the story" – a discovery that comes through interaction.

There are many examples of participatory music, across different cultures and through time. Turino (2008) discusses participatory music practices including Peruvian Aymara musicians, Zimbabwean Shona music and American Contra dancers, all of which involve the audience actively doing something during the performance. A number of musicians write pieces which foreground explicit audience participation. Frankfurt's Ensemble Modern work extensively with audience participation through their CONNECT <the audience as artist> series which invites composers to write new pieces with audience involvement. Electronic musicians Mark Fell and Rian Treanor have developed participatory pieces through their web-based *Intersymmetric* synthesiser. John Cage's *4'33"* could be seen to involve audience participation, or serving as a meta-commentary on what the audience brings to a piece. None of these pieces are explicitly aspiring to "immersion", but suggest ways in which audiences can interact with or participate in a musical work.

Whilst interaction and participation may facilitate immersion, there are critical discourses around these phenomena. Participatory art has come into criticism (see for instance Bishop, 2012 and Berry, 2015). Previous research by one of the authors of this chapter critically evaluated a series of participatory music projects, considering the pitfalls of participation (Tanaka and Parkinson, 2018). Shaughnessy (2012) argues that for all its participatory promise, immersive theatre can in fact be "coercive", the audiences policed, and the freedoms it offers merely "rhetorical". Some theatre critics have even accused immersive theatre of "low-level fascism" in the way audiences are manipulated whilst having the illusion of choice (White, 2012). The logistics of a performance mean audiences end up with a mere simulation of participation. As Biggin (2017, p. 11) neatly summarises: "The pragmatic necessities of immersive theatre get in the way of the promised pleasures of immersive experience." Perhaps helpful here is following Ryan (2001) in making a distinction between "weak" and "strong" forms of interactivity, the latter being a choice between predetermined alternatives, the former where the audience member plays an active role in the production of the work. The practicalities of safely and reliably staging a performance may require boundaries to be established around what can be allowed to happen. As a result of this, some interactive, immersive experiences tend towards weak interaction.

Christopher Small's (1998) concept of *musicking* offers a valuable, expanded perspective on participation. Small (1998, p. 9) proposes using "music" as a verb, writing: "To music is to take part, in any capacity, in a musical performance, whether by performing, by listening, by rehearsing or practising, by providing material for performance (what is called composing), or by dancing." This concept reminds us of the multitude of ways in which an audience member can participate in a musical event. There are resonances with Small's writing in theatre studies. In his essay "Gentle Acts of Removal", theatre maker Andy Smith notes that the sitting, listening theatre audience can be seen as participating and not merely passive. He goes further by stating: "Our approach to keeping this work open attempts to see us all as collaborators: co-directors, performers, technicians, front-of-house staff, audience" (Smith, 2011, p. 411). Simon Waters (2021) builds on Small's ideas and considers how musical instruments (and performances with them) can be designed to create "contexts for musicking". Performances that create space for social and technological interactions, as well as activities such as dancing, can create immersive audience experiences.

14.4.3 Insight 3: Immersion Is Multifaceted

Across disciplines, scholars identify the "multidimensionality of immersion" (Lee, 2020, p. 3 and Thon, 2008), recognising that it can describe a spectrum of experiences. One widely used model is Laura Ermi and Frans Mäyrä's (2005) SCI model for game immersion. This model identifies Sensory-based, Challenge-based and Imagination-based immersion. The SCI model has been adapted and developed, with some offering alternative categories or further granularity within these categories (Lee, 2020).

In the SCI model, Sensory-based immersion relates to the often impressive and even overwhelming audiovisual scale of gaming experiences which barrage the senses: "Large screens close to player's face and powerful sounds easily overpower the sensory information coming from the real world, and the player becomes entirely focused on the game world and its stimuli" (Ermi and Mäyrä, 2005, p. 7). Challenge-based immersion relates to the interactive element of games, and is achieved when there is a "satisfying balance of challenges and abilities" for the player, drawing on mental skills, motor skills or both (Ermi and Mäyrä, 2005, p. 7–8). Imagination-based immersion is that "in which one becomes absorbed with the stories and the world, or begins to feel for or identify with a game character" (Ermi and Mäyrä, 2005, p. 8).

As in gaming, there are different ways of being immersed in a musical performance, and different types of musical immersion that can coexist. Although alternative categories of immersion can be construed, the SCI model is a useful starting point, and exploring how to map these categories onto music may provide inspiration or compositional prompts for creating an immersive sonic experience. Sensory and challenge-based immersion will be covered here; imagination-based immersion has some overlap with narrative immersion discussed in 14.4.4.

Sensory-based sonic immersion may be achieved through playback technologies that literally immerse or envelop the audience in sound, blocking out other sensory stimuli or overwhelming them. This is what is conventionally understood as immersive audio, and may be achieved by playback systems such as d&b Soundscape, high order ambisonic sound systems or the BEAST system. Challenge-based sonic immersion suggests an alternative route to audience immersion, and could include music that requires such concentrated listening that the listener becomes immersed. The recent output of electronic duo Autechre, such as *NTS Sessions* (2018) is widely considered to be challenging, but fans find it intensely immersive, and the band tend to perform in dark rooms with no visuals to allow for maximum concentration and engagement with the sound. Fans describe the experience of listening to their complex beats as being "a real meditative immersion" (SmilingIvan, 2023). One review describes how "the listener feels immersed in a radical feeling of becoming" and another suggests a recent Autechre EP "offers an experience of immersion" (Schneider, 2016 and Gonsher, 2013). All of this perceived immersion emerges not from spatial sound but from qualities in the compositions and sounds themselves. Autechre are but one of many examples of "challenging" music, and the relationship between listening to such music and the types of psychological immersion it engenders in listeners may be a rich area for further research.

The SCI model and others are a reminder that immersion is a complex phenomena, describing a range of different experiences and psychological states. Immersive audio need not aspire to just one type of immersion, and exploring different types of audience immersion presents a rich territory for artistic research.

14.4.4   Insight 4: Immersion Through Narrative

Narrative elements may offer routes to psychological immersion. Narrative here refers not just to story but to the construction of the whole "world" of the experience. It can be related to the aforementioned imagination-based immersion identified by Ermi and Mäyrä but is also identified by Adams and Rollings (2006), Ryan (2001) and others.

The idea of being immersed in a narrative should be relatively familiar to anyone who has enjoyed a novel. This is, of course, an immersion that doesn't involve particularly new technologies. Immersive literary styles can be traced to the aesthetics of the nineteenth-century novel, with Dickens's works being a prime example (Ryan, 2001). Ryan (2001, p. 121) identifies three types of immersion that emerge from involvement with narrative: "spatial immersion, the response to setting; temporal immersion, the response to plot; and emotional immersion, the response to character." These categories of narrative immersion have been used in game studies: for instance, Thon (2008, p. 35) describes spatial immersion as "the player's shift of attention from his or her real environment to the game spaces" along with a mental construction of the game space.

It may be most tempting to connect this spatial immersion to spatialised audio, but Ryan (2001) relates "spatial immersion" to an evocation of place rather than a model of space, referring to the "madeleine effect" described by Marcel Proust, when a sensation can powerfully conjure up vivid mental images of a place distant in time or space. A single smell, taste or sound can be enough to virtually transport and immerse the listener in a distant place. Certain composers and musical aesthetics may be forever bound to certain places: it may be hard for some to hear the music of Ennio Morricone without being transported to the Mexico-US borderlands, or to hear Ralph Vaughn Williams's symphonies without imagining the southern English countryside. R. Murray Schafer (1994, p. 274) coined the term "soundmark", derived from landmark, to describe "a community sound which is unique or possesses qualities which make it specially regarded or noticed by the people in that community". Such soundmarks are so tied to place and community that they may also wield immersive power. Sound recordist Chris Watson has made numerous recordings that draw on this: *Stepping into the Dark* (1996) is recorded at sites including Kenya's Mara River and the Embleton Rookery in Northumberland, England. As Watson (1996) describes:

> Playing a recording made at one of these sites can recreate a detailed memory of the original event. Also, as others have described, there is an intangible sense of being in a special place — somewhere that has a spirit — a place that has an 'atmosphere'.

Ryan (2001, p. 123) also notes: "In the most complete form of spatial immersion, a sense of place is complemented by a model of space". Rich sonically immersive experiences may be achieved through a synthesis of sonic spatial evocation and spatial audio, in works that combine site-evocative field recordings and multichannel audio, such as Watson's *Namib* (2022).

Ryan (2001, p. 141) notes that "the phenomenological basis of temporal immersion and general [. . .] is a 'lived' or 'human' experience of time, as opposed to what may be called 'objective' or 'clock' time". Music can be highly effective at manipulating this human experience of time. One musical idiom that does this particularly well is long-form drone pieces by artists such as Eliane Radigue, Phill Niblock or Stephen O'Malley. In its eschewal of suspense or development, drone music perhaps provides the opposite of "narrative" in the storytelling sense, and a different experience of time to the Ryan's (2001) temporal immersion which is more concerned with the experience of suspense. Drone music can "obscure any sense of the passage of time" (Demers, 2010, p. 93), "suggest an alternative sense of time" (Demers, 2010, p. 99) and be an "immersive sacred sound" (Boon, 2002). Avante garde minimalist and drone composer La Monte Young describes a state of "listening in the present tense" evoked by drones (Boon, 2002), resonating with the idea of presence being equated to immersion in VR discussed in 14.3.4. Marcus Boon (2002) describes both temporal escape and an immersion in the sound itself provided by drone, noting: "Freed, at least temporarily, from the distraction of change and time, the listener enters the stream of the sound itself." Drones are just one way in which music manipulates the experience of time, and rhythmic or repetitive music can manipulate time in other ways, all of which may result in degrees of psychological immersion for an audience.

Finally, sound and music offer their own numerous routes to emotional immersion. It is widely acknowledged that listening to music can manipulate emotions, and research into the mechanisms of this is a whole field of study, and the experience of an overwhelming and powerful emotion could be said to be immersive (Juslin and Sloboda, 2011). Ryan's conception of emotional immersion refers to the emotional response to characters in textual worlds and as such, there may be connections to explore with *leitmotif*, the compositional technique which pairs musical themes with characters and is a common device in film music (Kassabian, 2001).

Literature and music can powerfully manipulate an audience's emotions as well as their perception of time and place in ways that are different and unique to the medium, and translating Ryan's concept across disciplines risks transforming it. Nonetheless, these concepts provide valuable starting points when thinking through compositional tools and sound design strategies for creating sonic immersion.

14.4.5 Insight 5: It's Not All About (New) Technology – Avoid Technology-Centric Approaches

Immersive forms should not wholly be understood in terms of new technologies, and one should be wary of reducing any immersive piece to the technologies involved. Immersion and technology have a complex relationship, and an immersive experience may imply the presence of some new cutting-edge digital technology. However, across disciplines, scholars express wariness about focusing too much on technologies.

Ermi and Mäyrä (2005) note that video games with basic (i.e. un-technologically advanced) graphics can still be immersive. Tetris is a good example of a game that is graphically and technologically unsophisticated relative to its contemporaries but is immersive for players. McMahan (2003) argues that big screens and high-fidelity sound are less essential than a well-constructed gaming world that matches a user's expectations.

Immersive heritage can create immersive experiences through use of space, inventive storytelling and audience interaction with little or no new technology; Kidd gives examples including Pollock's Toy Museum and the Imperial War Museum's *The 1916 Experience* and notes: "There are many analogue experiences that can be considered immersive: street games, interactive theatre and built environments such as theme parks and historic sites for example" (Gröppel-Wegener and Kidd, 2019). Writing recommendations for immersive experiences in museums and galleries, Kidd and McEvoy (2019) suggest, "avoid technology-centric approaches" advocating instead for a focus on storytelling. Regardless of the technologies involved "any and all heritage might *potentially* be understood as immersive" (Kidd, 2018).

As has been noted, immersive audio is often bound up with (relatively) new technologies. However, as Boren (2017) notes, *all* sound is inherently three-dimensional due to the way in which soundwaves propagate through air. As well as modern speaker playback technologies, older technologies can be used to create immersive musical experiences. In prehistory, it is possible that certain caves were used for gatherings and rituals because of their immersive acoustic properties (Boren, 2017). Rebelo (2021) also argues that the relationship between sound and space predates sound reproduction through multiple speakers and the ability to control sound diffusion, and goes beyond a purely technological or perceptual framework. He draws on musical traditions where the design of instruments and their position within an environment provide examples of a hidden connection contract between sound and acoustical space, tracing the evolution of Western classical music and its relationship to religious architecture. For instance, the fifteenth-century practice of *cori spezzati* achieved an immersive spatialisation of sound by distributing multiple choirs through a chapel.

Some contemporary artists spatialise sound in inventive ways that don't follow traditional ideas of multichannel sound. The performance collective Katz Mulk wander through the audience with portable amplifiers and radios to spatialise the sound. Annea Lockwood's *Three Short Stories and an Apotheosis* (1985) uses a ball with embedded speakers in it that is passed through the audience, introducing both a spatial and participatory element. Gordon Monahan's (1987) *Speaker Swinging* explores the musical affordances of a moving speaker, adding spatial movement to the sound but also introducing acoustic effects such as the Doppler effect. These two pieces are discussed further in Van Eck (2017) which explores extensively the idea of speakers as instruments. Stockhausen's *Helikopter-Streichquartett* (1995) puts string players in helicopters that fly over the audience, though this is far from the most practical or environmentally friendly way to immerse an audience in spatialised sound.

As these examples demonstrate, immersive audio, in the widely understood definition of sound that envelops the audience, is something that can be realised without the standard immersive playback and spatialisation systems.

14.5 Discussion: Immersed In Immersion

Having reflected on insights into immersion from different disciplines and how this could inform sonic practices, for the final part of this chapter these immersive practices will be contextualised within what is termed the "immersive economy", considering how this broader context may impact practices. "Immersive industries" or the "immersive economy" are used to describe companies working with emerging technologies of Virtual Reality (VR), Augmented Reality (AR), Mixed Reality (MR) and Extended Reality (XR).

In 2023, the creative industries are truly immersed in immersion. Google Trends results for "immersive" shows a continual growth in popularity since 2013. The past few years have seen a number of large-scale immersive events, in part driven by the COVID-19 pandemic. Alongside the aforementioned *Van Gogh* and *David Hockney* immersive exhibitions from recent months, Cannes film festival now has *Cannes XR* to platform and showcase immersive projects, and *Cannes XR 2023* saw the debut of a new VR Banksy experience. Since 2021, the *BFI London Film Festival* has hosted LFF Expanded to showcase immersive projects.

Within this context, there is a lot of excitement surrounding immersive audio. Nuno Fonseca's (2020) *All You Need to Know about 3D Audio* asserts that "Immersive Sound is the next big thing in audio". Roginska and Geluso's (2018, p. xii) edited collection *Immersive Sound: the Art and*

Science of Binaural and Multi-Channel Audio (2018) notes how "immersive sound is strategically aligned with the future of communication and entertainment".

In the UK, there has been significant UK Government funding for the Immersive Economy in the background of this growth in visibility of immersive cultural experiences, including:

- The *Creative Industries Cluster Programme* which distributed £56 m through the AHRC.
- The *Audience of the Future Challenge* which invested £39.3 m from 2018–2022.
- *CoSTAR*, a £69 m UKRI immersive tech programme
- *XR Stories* which saw £15 million investment from the University of York, the BFI, the AHRC, ERDF and Screen Yorkshire (Immerse UK, 2022).

A recently advertised AHRC call worth £6 m aimed to support "immersive innovations", "immersive technologies", "immersive production", "immersive content", "immersive projects" and "immersive industries" (UKRI, 2023).

The growth of the Immersive industries in the UK has been bound up with the UK Government's strategic planning, as evidenced in their Creative Industries Sector Deal and Industrial Strategy. The UK Government may be motivated to invest in this Immersive Economy because it shows potential of becoming a significant area of growth and for funding in the creative industries in the UK. The number of businesses working in immersive tech in the UK has increased 83% in the past 5 years and in 2021 had a combined turnover of £1.4 bn. More growth is predicted: PricewaterhouseCoopers (PwC, 2019) estimate that immersive technology will add $1.5 trillion to the global economy by 2030, with a $69.3 bn boost to the UK economy and enhancement of over 400,000 UK jobs. The immersive industries therefore represent an attractive area for state funding, because alongside growth there is an influx of private investment following state support: in 2021, the UK Immersive Industries received £224 million in private investment.

An important body in the UK's Immersive Industries is Immerse UK, an organisation established via an Innovate UK KTN (Knowledge Transfer Network) in 2016 to catalyse developments in the UK "immersive tech sector". Their founding partners include the Royal Shakespeare Company, the National Theatre, BAE systems, Unity, Epic Games, Alder Hey Children's Hospital and the University of York. Immersion, then, provides the impetus for an unlikely alliance between theatre, academia, the arms industry, gaming and the NHS.

Immerse UK has a technology-focused perspective on immersion. The *2022 Immersive Economy Report* presents a highly technology-centric view on the immersive industries, with no critical evaluation of immersive experiences or separation of form from experience. The concerns of Immerse UK are industrial, not artistic: the Immerse UK website asserts that their intention is to become "owned by industry for industry". Growth of immersive companies and the uptake of immersive technologies are their main concerns, rather than any interrogation of what makes an experience immersive.

One potential effect of this is that those developing immersive experiences may inherit the same technology-focused view of immersion, and immersive experiences (which may or may not use the technologies at the heart of the immersive industries) become synonymous with the technology. As has been argued, immersive experiences should be detached from these immersive forms.

Also worth noting is that the rise of immersion, and the generous funding available for immersive industries, takes place against a backdrop of significant cuts to arts funding in the UK. Treasury funding to the Arts Council per person decreased from £100 in 2007/2008 to £60 in 2017/2018. Many major UK arts organisations, including English National Opera, have in recent years lost

100% of their guaranteed Arts Council funding. Cultural funding has decreased 46% since 2005 (Khomami, 2022).

For artists working with technology, gravitating towards "immersive" works looks like quite a sensible option in order to apply for the limited remaining arts funding available. Reframing one's creative practice as "immersive" could be a survival strategy. Immersive tech remains a safe area for government investment in the arts where art, at least potentially, makes money, happening within the milieu of start-ups and SMEs, and attracts private investment. However, there is a risk that creative work in this area becomes inseparable from the cutting edge of technology, and alienated from the experience of the audience.

14.6   Conclusion

This chapter has attempted to bring diverse approaches to audience immersion to the field of immersive audio, whilst contextualising the current interest in immersive technology within the broader "immersive industries". This chapter marks the beginning of a larger research agenda, and only scratches the surface in terms of processing the broad range of approaches to immersion and immersive experiences. There are of course far more insights to be taken from other disciplines where immersive experiences are found. An exploration of the essential social aspects of music, as well as the role of activities such as dancing, has been beyond the scope of his chapter. Nonetheless, the initial stages of this research have been fruitful in terms of helping us reflect upon own practices, and given us some next steps to develop these practices.

One of the many things not done in this chapter is to question whether immersion is even desirable in music. Critiques of immersion note that it is a barrier to critical thinking (Ryan, 2001). Future critical work may be to unpick some of the discourses that unquestioningly accept the idea that musical experiences should be immersive, and even explore how immersive forms can be used to create intentionally non-immersive experiences.

Finally, we would like to revisit our earlier paraphrasing of Kidd: any sonic experience has the potential to be immersive, and each time we find ourselves having had an immersive sound experience, we will be critically reflecting on the rich tapestry of factors that enabled it.

References

Adams, E. and Rollings, A. (2006). *Fundamentals of Game Design*, San Francisco, Peachpit.

Agrawal, S., Simon, A.M.D., Bech, S., Bærentsen, K.B. and Forchammer, S. (2020). Defining Immersion: Literature Review and Implications for Research on Audiovisual Experiences, *Journal of the Audio Engineering Society*, Vol. 68, No. 6, pp. 404–417.

Berry, J. (2015). Everyone is Not an Artist: Autonomous Art Meets the Neoliberal City, *New Formations*, Vol. 84, No. 84, pp. 29–39.

Biggin, R. (2017). *Immersive Theatre and Audience Experience*, Basingstoke, Palgrave Macmillan.

Bishop, C. (2012). *Artificial Hells: Participatory Art and the Politics of Spectatorship*, London, Verso.

Boon, M. (2002). The Eternal Drone, in Young, R. (ed.), *Undercurrents: The Hidden Wiring of Modern Music*, London, Bloomsbury.

Boren, B. (2017). A History of 3D Sound, in Roginska, A. and Geluso, P. (eds.), *Immersive Sound: The Art and Science of Binaural and Multi-Channel Audio*, Abingdon, Routledge.

Brown, E. and Cairns, P. (2004). A Grounded Investigation of Game Immersion, *CHI'04 Extended Abstracts on Human Factors in Computing Systems*, Vienna, pp. 1297–1300.

Bucher, J. (2017). *Storytelling for Virtual Reality: Methods and Principles for Crafting Immersive Narratives.* Abingdon, Routledge.

Collins, K. (2013) *Playing with Sound: A Theory of Interacting with Sound and Music in Video Games*, Cambridge, MIT Press.

Demers, J. (2010). *Listening Through the Noise: The Aesthetics of Experimental Electronic Music*, Oxford, Oxford University Press.

Ermi, L. and Mäyrä, F. (2005). Fundamental components of the gameplay experience: Analysing immersion, *Proceedings of the 2005 DiGRA International Conference*, Vancouver.

Fonseca, N. (2020). All You Need to Know about 3D Audio (website), available online at https://soundparticles.com/resources/ebooks/3daudio/ [accessed August 2023].

Gonsher, A. (2013). Autechre L-Event EP (website), *XLR8R*, available online at https://xlr8r.com/reviews/l-event-ep/ [accessed January 2024]/.

Gröppel-Wegener, A. and Kidd, J. (2019). The Seductive Power of Immersion, in Gröppel-Wegener, A. and Kidd, J (eds.), Critical Encounters with Immersive Storytelling, Abingdon, Routledge.

Immerse UK. (2022). The 2022 UK Immersive Economy Report (website), available online at www.immerseuk.org/resources/the-2022-uk-immersive-economy-report/ [accessed July 2023].

Jennett, C., Cox, A. L., Cairns, P., Dhoparee, S., Epps, A., Tijs, T. and Walton, A. (2008). Measuring and Defining the Experience of Immersion in Games, *International Journal of Human-Computer Studies*, Vol. 66, No. 9, pp.641–661.

Juslin, P. N. and Sloboda, J. (eds.). (2011). *Handbook of Music and Emotion: Theory, Research, Applications*, Oxford, Oxford University Press.

Kassabian, A. (2001). *Hearing Film*, Abingdon, Routledge.

Khomami, N. (2022). Arts Funding Cuts 'Cultural Vandalism', Says Juliet Stevenson, at DCMS Protest, *The Guardian*, 22nd November, available online at www.theguardian.com/culture/2022/nov/22/arts-council-england-cuts-are-cultural-vandalism-says-juliet-stevenson [accessed August 2023].

Kidd, J. (2018). 'Immersive' Heritage Encounters, *The Museum Review*, Vol. 3, No. 1.

Kidd, J. and McEvoy, E. (2019). Immersive Experiences in Museums, Galleries and Heritage Sites: A Review of Research Findings and Issues, *Creative Industries Policy and Evidence Centre*, available online at https://pec.ac.uk/discussion-papers/immersive-experiences-in-museums-galleries-and-heritage-sites-a-review-of-research-findings-and-issues [accessed August 2023].

Kim, S. (2021). Contextual Factors in Judging Auditory Immersion, in Paterson, J. and Lee, H. (eds.), *3D Audio*, pp. 160–174. Abingdon, Routledge.

Lee, H. (2020). A Conceptual Model of Immersive Experience in Extended Reality, PsyArXiv. https://osf.io/preprints/psyarxiv/sefkh

Lombard, M. and Ditton, T. (1997). At the Heart of It All: The Concept of Presence, *Journal of Computer-Mediated Communication*, Vol. 3, No. 2.

McMahan, A. (2003). Immersion, Engagement, and Presence: A Method for Analyzing 3-D Video Games, in Wolf, M. and Perron, B. (eds.), *The Video Game Theory Reader*, pp. 67–86. Abingdon, Routledge.

PwC. (2019). Seeing Is Believing: How VR and AR Will Transform Business and the Economy (website), available online at www.pwc.com/gx/en/industries/technology/publications/economic-impact-of-vr-ar.html [accessed August 2023].

Rebelo, P. (2021). Sound and Space: Learning from Artistic Practice, in Paterson, J and Lee, H. (eds.), *3D Audio*, Abingdon, Routledge.

Roginska, A. and Geluso, P. (eds.). (2018). *Immersive Sound: The Art and Science of Binaural and Multi-Channel Audio*, Abingdon, Routledge.

Ryan, M. L. (2001). *Narrative as Virtual Reality. Immersion and Interactivity in Literature*, Baltimore, Johns Hopkins University Press.

Schafer, R. (1994). *The Soundscape: Our Sonic Environment and the Tuning of the World*, Vermont, Destiny Books.

Schneider, B. (2016). AUTECHRE: ELSEQ 1-5 (website). *Caesura*, available online at https://caesuramag.org/posts/autechre-elseq-1-5 [accessed January 2024].

Shaughnessy, R. (2012). Immersive Performance, Shakespeare's Globe, and the 'Emancipated Spectator', *The Hare*, Vol. 1, No. 1.

Small, C. (1998). *Musicking: The Meanings of Performing and Listening*, Middleton, Wesleyan University Press.

SmilingIvan. (2023). Wow! Never Heard of Autechre. Now Listening to a Few Songs. Wow! *Reddit*, available online at www.reddit.com/r/aphextwin/comments/15biq83/wow_never_heard_of_autechre_now_listeni ng_to_a/?rdt=59529 [accessed January 2024].

Smith, A. (2011). Gentle Acts of Removal, Replacement and Reduction: Considering the Audience in Co-Directing the Work of Tim Crouch, *Contemporary Theatre Review*, Vol. 21, No. 4, pp. 410–415.

Sterling, C. (2020). Designing 'Critical' Heritage Experiences: Immersion, Enchantment and Autonomy, *Archaeology International*, Vol. 22, No. 1, pp. 100–113.

Tanaka, A. and Parkinson, A. (2018). The Problems with Participation, in Emmerson, S. (ed.), *The Routledge Research Companion to Electronic Music: Reaching out with Technology*, Abingdon, Routledge.

Thon, J. N. (2008). Immersion Revisited: On the Value of a Contested Concept, in Leino, O., Wirman, H. and Fernandez, A. (eds.), *Extending Experiences: Structure, Analysis and Design of Computer Game Player Experience*, Lapland, Lapland University Press.

Torino, T. (2008). *Music as Social Life: The Politics of Participation*, Chicago, University of Chicago Press.

UKRI. (2023). XRtists: Supporting the Implementation of Immersive Technologies (website), available online at www.ukri.org/opportunity/xrtists-supporting-the-implementation-of-immersive-technologies/ [accessed August 2023].

Van Eck, C. (2017). *Between Air and Electricity: Microphones and Loudspeakers as Musical Instruments*, London, Bloomsbury Academic.

Van Elferen, I. (2016). Analyzing Game Musical Immersion: The ALI Model, in Kamp, M., Summers, T. and Sweeney, M. (eds.), *Ludomusicology: Approaches to Video Game Music*, Sheffield, Equinox Publishing.

Waters, S. (2021). The Entanglements Which Make Instruments Musical: Rediscovering Sociality, *Journal of New Music Research*, Vol. 50, No. 2, pp 133–146.

Watson, C. (1996). *Stepping into the Dark (liner notes)*, available online at https://chriswatsonreleases.bandc amp.com/album/stepping-into-the-dark [accessed January 2024].

White, G. (2012). On Immersive Theatre, *Theatre Research International*, Vol. 37, No. 3, pp. 221–235.

Wilkinson, M., Brantley, S. and Feng, J. (2021). A mini review of presence and immersion in virtual reality, *Proceedings of the Human Factors and Ergonomics Society Annual Meeting*, Baltimore, Maryland, Vol. 65, No. 1, pp. 1099–1103.

Discography and Music References

Autechre (2018), [CD] *NTS Sessions*, London: Warp Records.

Lockwood, Annea (1985) [score] *Three Short Stories and an Apotheosis*.

Monahan, Gordon (1987), [digital download] *Speaker Swinging*.

Stockhausen, Karlheinz (1995), *Helikopter-Streichquartett*, Holland Festival, 26 June 1995.

Watson, Chris (1996), [CD] *Stepping into the Dark*, London: Touch.

Watson, Chris (2022), [sound installation] *Namib*, Indexical, San Francisco.

15

DELIBERATE PRACTICE AND UNINTENDED CONSEQUENCES IN MUSIC PRODUCTION AS PRACTICE AND PEDAGOGY

Hussein Boon

15.1 Introduction

This chapter continues my ongoing exploration of creative processes and their application to the DAW (Digital Audio Workstation). I focus on exploring interesting excursions within the DAW and articulating suitable concepts and ideas with broad appeal and application in music production and teaching. Many of these concepts can also be applied to other aspects of musical work, such as songwriting, instrumental performance and sound design. I have also shown how ideas drawn from neo-Riemannian Operations can be used and taught within the DAW (Boon, 2024, pp. 308–316). For those who self-produce, many will find these ideas beneficial, especially if looking for challenges to go beyond tools and genre-based thinking. I draw from practices including dub and Sawari (Takemitsu, 1995), and, as a binding principle central to the ideas and work discussed in this chapter, I also deploy part of Peter Osborne's art schema (Osborne, 2013). Osborne is primarily used as an umbrella term for the more usual idiomatic terms such as creative abuse (Keep, 2005), glitch and failure (Cascone, 2000), error (Baxter, 2019), unintended consequences (Théberge, 2001; Zagorski-Thomas, 2014), as well as concepts of creative constraints, such as in Eno's Oblique Strategies (Boon, 2021, p. 311; Eno Shop, 2010; McNamee, 2009).

This chapter is divided broadly into three sections. The first deals with various positions and discussions concerning ideas of 'you're not supposed to do that'. The second seeks to draw these multiple strands under the umbrella of Osborne's schema. I have previously used Osborne's schema in the context of sampling (Boon, 2023e), and this current chapter contributes to the broader topic of music production. To reiterate, Osborne provides a more neutral language basis to support exploration rather than that previously afforded by terms such as creative abuse and failure. Whilst these terms might appeal to some, they also might not find acceptance among all practitioners, especially in the teaching context. Third, I provide several examples from my practice and explorations within the DAW, supported by video.

15.2 'You're Not Supposed To Do That' and Other Subverting Terms

The idea 'you're not supposed to do that', at least from my perspective, can be said to sit within several creative music areas, particularly those widely acknowledged as having a basis in subverting

DOI: 10.4324/9781003396710-15

[production] practices. Among these include examples like the all buttons in mode of the 1176 compressor (Moore, 2019; Zagorski-Thomas, 2014, p. 129) to DJ Kool Herc's ground-breaking twin decks approach (Ewoodzie Jr., 2017, p. 4; p. 41). In the music literature, these practices are often declared as matters of "creative abuse"(Keep, 2005), glitch and/or failure (Cascone, 2000).

For example, Barker suggests that "the potential for error marks the potential for the new and the unforeseen, we can see that an error in itself may be creative. An error may be utilised" (Barker, 2007). Whilst errors can be useful, an error cannot be assumed to be creative in and of itself. The practitioner must be able to recognize its potential, even if dimly.

Théberge clarifies the role of this terminology and identifying these approaches as "instrumental in defining a particular 'sound'" (Théberge, 2001, p. 4). Therefore, new sound can be considered constituted at the individual sound/sample level. However, it can also mean the total sound, such as the sound of a production or piece, or a product of new performance techniques.

15.2.1 Intentional and Unintentional Actions

Various researchers have identified the role of intentional and unintentional actions in establishing new approaches to sound. Cascone highlights that "new techniques are often discovered by accident or by the failure of an intended technique or experiment" (Cascone, 2000, p. 13). Others suggest that "failure could be a good source for change" (Trio Sawari, 2017, p. 412). Yet, new sound can also be found in accidental encounters with instruments and/or technology, such as Gary Numan and the Minimoog "synthesizer in the corner" (Numan, 2020, p. 46), left behind from a previous studio session. Thus, ideas of 'you're not supposed to do that' can also mean adapting the musical spikiness of punk fixed into electronic instruments as Numan did (2020, p. 46) as much as it can mean the exploitation of glitchy artefacts from experimental or failed processes. Therefore, the contribution of not doing things in the prescribed manner is a prerequisite for much music-making practice in the broadest sense. Yet, despite the importance of these various positions and/or approaches, locating these in a teaching practice, curriculum or pedagogy may also be challenging.

An important point to consider concerning ideas of failure, unintentionality, and happy accidents is that whilst they might trigger a new idea, these discoveries, once understood, are usually refined more deliberately. This idea of a deliberate practice can also be extended to AI. YACHT, when working with AI using Runway ML (2023) and Google's Magenta (n.d.) – trained on their material – capitalized on the glitch and generative failure produced by the machine and saw this as a challenge for their work. In this context, the band interpreted the AI as failing to comprehend what they were looking for in a melody. Still, because of this "failing at doing exactly what they're supposed to do" (Mattise, 2019), the results "tend to fall just outside the boundary of what's an acceptable, normal, or traditional melody" (Mattise, 2019). Thus, the band used this failure as a point of creativity for further human development rather than reject the machine outputs as wholly unusable or attempt to use them without modification.

15.2.2 The Flaw In Subverting Terms

With subverting terms like creative abuse and failure, many of these also have a flaw. That is, their very wording might not have a general appeal and, therefore, act as a widespread barrier to their adoption. Their general utility is important because, if one agrees with Théberge's point, then in the practice of music, these approaches may/do play a (significant) role in the development of the new. Yet, terms like creative abuse may lack a general appeal due to the difficulty of language or that (some) practitioners may feel that this type of approach is reserved for more esoteric music genres. In that case, it might also be challenging to identify a suitable point in a curriculum to incorporate such a focus. For example, consider the shift in control mechanisms such as Dynamic Range Compression (DRC) shifting from a preventative measure when tracking to tape to a more creative tool that imparts coloration to recorded material (Moore, 2019, p. 212).

Subverting approaches and rule-breaking – even if mildly – can also be considered a form of deregulation. Moore's discussion and examples of coloration, as well as Kerridge's description of going into the red at recording sessions, learned from Joe Meeks (Kerridge, 2016, p. 59), can, at least in the first instance, be considered as examples of irregularity and deregulation, until they are adopted as practice and, therefore, become conventions. These examples can be thought of as "potential irregularities [which] can be fed back into the system without unduly disturbing it" (James, 2014, p. 143). Therefore, ideas of creative abuse or unintended consequences need not always result in extremes; they can also be *soft ruptures*. This also reveals why terms like creative abuse, whilst they aim to stimulate exploration, also communicate ideas that might be interpreted as brutal and destructive when learning new and potentially divergent (production) skills.

Baxter acknowledges the role of experimentation inherent in the "bedroom production route" (Baxter, 2019, p. 287) and the role this may have in a taught course. However, this chapter argues that experimentation in relation to producing creative outputs is neither an adjunct nor inferior to a technical understanding. Where developing technical understanding is described as becoming "almost second nature" (Baxter, 2019, pp. 286–287), the same can be said of creativity, but with a critical difference. A technical understanding is not sufficient to drive creative work alone. Furthermore, "skill is often an ideological category imposed on certain types of work by virtue of the sex and power of the workers who perform it" (Phillips and Taylor, 1980, p. 79). As much as second nature might be the goal of a skilled technical understanding, the artistic sensibility implied in the bedroom route needs to be amplified rather than being pushed to one side when pursuing technical matters.

15.2.3 Dub Practice As Deregulation

One such practice that can be considered as meeting James' ideas of deregulation is dub. As a practice, dub does not destabilize the system of music production. In fact, it enhances and expands it by giving more to the field of production, not just concerning production approaches – which include the role of bass, sub bass and effects (such as dub delay) – but also additional practices such as remixing and recycling. These artefacts and practices are still evident in music production today. In discussing King Tubby, Williams highlights that "Tubby's knowledge and expertise with the technology at his disposal not only shaped his music but also influenced the many different genres and styles of recorded music making across the world" (Williams, 2012, p. 235–236). The practice of preparing the mixing desk for performance, rapidly muting channels, and controlling effects that are on the edge of descending into uncontrollable feedback could be described as a pretty chaotic approach to mixing. As such, all of these practices can be thought of as aspects of deregulation. Dub deregulates what a mixing desk is and what is possible with recorded material. In doing so, dub practitioners reveal new insights drawn from recorded material and generate new flows and directions in what is possible with recorded material mediated via the mixing desk, which also translates into general production practice.

When talking about the dub mixing approach, producer and reggae and dub pioneer Dennis Bovell described preparing mixing desks for a dub session as "generally just have fun from the engineer's point of view" (Bovell, 2017). Yet dub can also experience its own deregulation. In an interview in 2017, Paul Gilroy, when discussing Dennis Bovell and dub, said:

> [Bovell's] ideas of dub didn't come from Jamaica directly, but they were born from listening to the experimental music that Hendrix was making in that they [Bovell's ideas] were connected to that.
>
> *(Gilroy, 2017)*

While Hendrix might seem an interesting and less well-documented participant in the story and development of dub, he plays a role in the development of Bovell's dubwise imagination as a departure from purely Jamaican influences.

15.2.4 Sawari, Inconvenience and Dub Practice

Another approach for consideration is the Japanese practice of *Sawari*. This can be considered as utilizing deliberate inconvenience to create sound (Takemitsu, 1995, p. 65). As an immediate example, reconfiguring a mixing desk and multitrack recordings in such a radical fashion, as in dub mixing or no-input mixing (Chamberlain, 2018), can be considered a type of inconvenience. Not only does it facilitate skill in planning and managing, it also and builds a production identity to develop a sound (Bovell, 1979, p. 7). The purpose of inconvenience can be considered relative to other practice terms, such as creative constraints (Boon, 2021, p. 319). Both inconvenience and creative constraints function with a similar goal in mind: the production of new work.

Pat Metheny describes the ease with which guitarists, and guitar playing, can get "sucked into the web of fingers, patterns, positions and grips" (Metheny, 1985). He gives one example of how he challenges himself, which can be interpreted as either constraint or inconvenience, by using a twelve-string guitar where he

> put all kinds of weird strings on it and give it to my girlfriend to tune it until she thought it sounded right. I'd then try to get something out of it, maybe making a modification here and there.
>
> *(Metheny, 1985)*

It is interesting to note that Metheny's approach, certainly in this example, is generally in the domain of tools, specifically around retuning, restructuring and organization of the guitar. His thinking tools, internal logic and technique remain largely untouched, as he says he always thinks in interval relationships (Metheny, 1985). Encapsulated within Sawari is the sense that new techniques might be required to produce new sound, which will be awkward and/or inconvenient to the practitioner. A mixing desk, transformed for dub mixing, remains inconvenient *because* of the fixed nature of the desk. Dub, therefore, is perhaps less of a constraint-based activity and perhaps closer to Sawari in its conditions of inconvenience.

15.2.5 Example 1 – Smear Symphony

Reflecting on this in my work, the following account shows how deliberate inconvenience might be applied in a DAW. In the example, *Smear Symphony* (Boon, 2022), I devised several processes. Firstly, I used a vinyl click distortion plugin to generate noise. The vinyl plugin allows for the density of clicks and their dispersion, i.e. frequency range, to be controlled.

TABLE 15.1 Resonator pitches

Primary Pitch	b3	5	2	b7
0	+3	+7	+2	-2
C	Eb	G	D	Bb

This is particularly useful in controlling the overall density of the piece. The output from the vinyl plugin is then modified through other processors, such as resonators and large reverbs, to produce the sound piece.

The resonators are used to structure the unpitched outputs from the vinyl plugin into an overall pitch structure of Cm9. The values shown in Table 15.1 have the M2nd (which I treat as the 9th) and b7th, with both close voiced on either side of the root. This allows for a type of tension to arise in the piece via this clustering of secundal pitches. I use two resonators for this piece, and whilst both are organized using the same pitch collection, one can also be played using MIDI controller messages. This allows various pitch settings to be explored within the pitch collection, from single notes and various dyadic and triadic collections to the full Cm9 chord.

This piece is further shaped by the use of the tablet-based interface, using Liine's Lemur (Boon, 2021, p. 317), which turns these processes into a performable production. As Hugill points out, "digital musicians are always responsible for building their own instruments, in some cases from preassembled components" (Hugill, 2012, p. 167). As such, the piece and the tablet interface are tied together into what can be described broadly as a digital instrument.

In this setting, the DAW is treated as something to be performed with. In fact, sound can only be produced by playing and interacting with the plugins using the mouse, trackpad and/or tablet interface, but not by pressing play!

15.2.6 Osborne's Anti-Aestheticist Use of Aesthetic Materials

To complete this section, I discuss one additional aspect to describe much of this work in my practice. I use it primarily to replace phrases like creative abuse, glitch and failure. In many respects, what is being referred to in these various activities is what Peter Osborne describes as "the critical necessity of an anti-aestheticist use of aesthetic materials" (Osborne, 2013, p. 28). Whilst I have identified one part of Osborne's schema, the other five are equally useful but not necessarily pertinent to the current discussion.

The anti-aestheticist use of aesthetic materials, as well as being a more apt identification and summation of terms like creative abuse, is also flexible to encompass Takemitsu's Sawari, and dub mixing approaches. As a term, the anti-aestheticist use of aesthetic materials enables and promotes (accidental) discovery through the exploration of materials and/or their refinement as part of a deliberate practice. If one considers the mixing desk as an aesthetic material, then dub mixing expands what a mixing desk is and can be. Approaches like no-input mixing further challenge what a mixer is for and can be. Both dub and no-input mixing create conditions ripe for accidental discovery and unintended consequences either as novel instances or developed into deliberate processes. These can then be brought back into the more conventional understanding of mixing and mixers.

15.3 Examples

In this section, I discuss some examples of applying these ideas in the DAW Ableton, using their utility plugin known as the scale device, which I also refer to interchangeably as a quantizer.

15.3.1 About The Scale Device/Quantizer

The Ableton scale device consists of a 12x12 grid, shown in Figure 15.1. Incoming notes on the x-axis can be quantized to particular pitches on the y-axis. The incoming stream of notes might be found in a specific scale, such as a major scale, which can be changed to another scale, such as a minor pentatonic scale. The device is generally presented as beneficial for those with less secure keyboard-playing skills, where it can be used to correct incoming notes. The scale device named 1a (Figure 15.1) shows the quantizer in its default setting, which is the chromatic scale. This produces a 1:1 correspondence with input pitches matched at the output.

Scale device 1b (Figure 15.1) shows all incoming notes mapped to note G. This capability to remap incoming notes is developed more musically in the scale device labelled 1c (Figure 15.1), where notes are reconfigured to form a melodic figure which does not correspond to a scale. Furthermore, this implementation of the scale device also shows how, as a quantizer, it could be explored and performed with (Boon, 2023b), which transforms the scale device from a static, always-on utility tool into something more dynamic.

In the following sections, I show the scale device used in two creative applications, used more as a design tool rather than as a corrective processor.

15.3.2 Using Follow Actions and Clip-Based Automation

This approach consists of three notable elements (see video Boon, 2023c):

1. A set of contiguous MIDI clips
2. A collection of scale devices in a rack
3. Clip-based automation controlled by follow actions

The MIDI clips (Figure 15.2 at A) contain the same single note (Figure 15.2 at C). This provides the rhythmic drive for the part, which is in sixteenth notes. If different rhythmic values are required, more clips can be adapted to express new rhythmic values. The MIDI clips use follow actions (Ableton, n.d.; Boon, 2024, p. 315–316), which are set using the launch control window shown in Figure 15.2 at position B and in Figure 15.3.

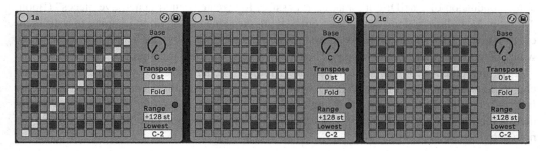

FIGURE 15.1 Three configurations of the Ableton scale device.

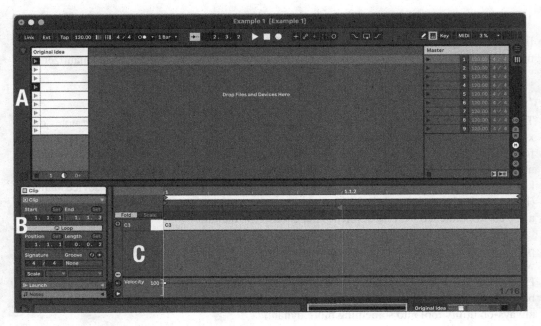

FIGURE 15.2 Follow actions and automation.

FIGURE 15.3 Follow action window (left) and conditions (right).

Follow actions are a type of branching logic with instructions for Ableton to perform an action at the conclusion of a MIDI clip, making it possible to "create pseudo-generative pieces" (Boon, 2023a, p. 3). Follow actions can range from playing the same clip again to jumping to a different one (Figure 15.3). The outcomes for these follow actions are set using the launch control window, which determines what should happen when a clip meets a particular condition.

Each instrument rack channel is named after the note being played by the scale device. Each clip contains automation (Figure 15.2, C), which re-positions the chain selector (Figure 15.4) of the instrument rack, which is mapped to a single macro, labelled *Chain Selector* (Figure 15.4). A MIDI note is required as a trigger, but the scale device and macro position combination determine the note output.

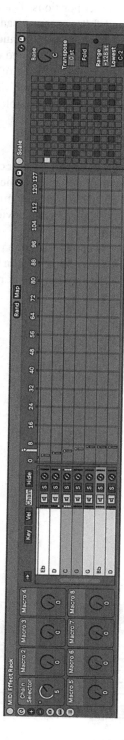

FIGURE 15.4 Instrument rack.

The instrument rack channels (Figure 15.4) are organized so that at some points, only a single note will play, and a chord will be heard for other positions. This demonstrates a certain versatility in how the combination of scale device and follow actions can be used to design a specific set of tonalities for a recording or songwriting session. Whilst much of this project is designed i.e. deliberate, at the point of execution, the outputs are determined by the follow actions, which may produce unintended consequences. This demonstrates a type of controlled randomization and/or generativity, which creates potential conditions for new output variations. Outputs can be captured as MIDI or audio for further editing and arranging into a complete piece.

15.3.3 Using A Performable Macro Control

The second example uses the instrument rack and chain selector controlled using a macro described in the previous example. However, rather than automating the macro using follow actions, the producer performs with it (see video Boon, 2023d). A randomization device (Figure 15.5a) also varies the incoming notes for the various configured scale devices. Its chance parameter is set to 100%, so that a new note is always produced.

Instead of single notes, each scale device is configured for a particular harmony (Figure 15.5b). The harmonies used in this example are Cm, G, Gm, G#dim and Am. Each macro position is available using the numeric keypad 0–4, though they can also be mapped to other controller devices.

FIGURE 15.5a The Ableton randomizer.

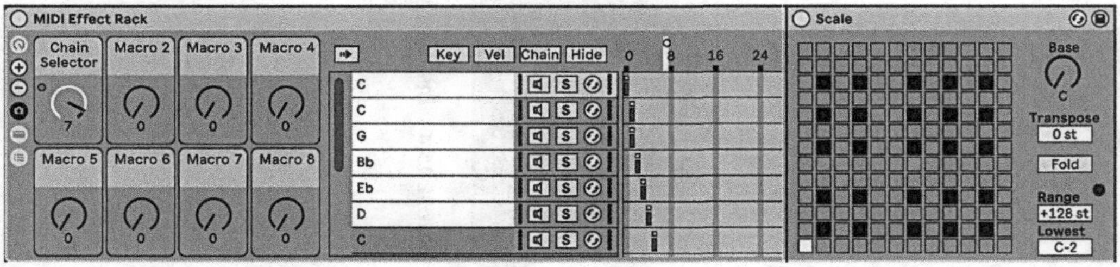

FIGURE 15.5b Ableton quantizer rack and macro.

There are some advantages to using this approach. Firstly, the producer can identify a chord palette to explore new combinations. Depending on how the scale device is organized, different outcomes that control the note density of the chord can be achieved. For example, the number of notes used for the diminished chord can vary from a single note to the complete triad. Secondly, as this is performed with, rather than automated, the producer has a great deal of control to identify interesting sequences of arpeggiated chords and prepare them as part of their songwriting and/ or arranging process. As chords are accessed by entering numbers directly from the keypad, the underlying patterns can be explored freely. Lastly, whilst the rack's organization suggests one potential direction of travel, 0–4, this is by no means the only available option. To only perform in a linear direction of travel is to miss some interesting chordal propositions.

15.4 Discussion

The examples used in this chapter demonstrate several ideas helpful in developing creative approaches using a DAW. The first example generates a piece from a process, and the second and third examples are more suited to generating parts. All conform to what Birch and Muniesa call an asset, which can be "owned, or controlled, traded, and capitalized as a revenue stream" (Birch and Muniesa, 2020, p. 2).

The first example adopts the approach of performance in production (Boon, 2023a, pp. 3–5; Moorefield, 2010, pp. 102–103), which means that production can be performed at live events. The performance can be treated as an asset, especially as live music is one of the few remaining domains where an artist or producer can play their work and earn some money. Therefore, this performance-production approach is an avenue that music production teaching should explore. The example piece, *Smear Symphony* (Boon, 2022), can be performed at a gig, at a club or as part of an installation, yet requires no specific adaptations or accommodations, meaning it can be played anywhere. Its length is variable, and the sense of what the piece can be can continue to be developed over time, from one performance to another.

There is also another reason for advocating this performance in production approach. In an age of AI and machines copying musical works, a musical piece which is always varied is more difficult for an AI company to reproduce. Therefore, performance in production, and vice versa, forms part of a resistance approach to the encroachment of AI in all creative areas. Morreale highlights that there is an unchallenged "axiomatic assumption that 'AI will help musicians make music'" (Morreale, 2021, p. 109) and that "much of the work conducted by AI companies is not to solve genuine artistic problems" (Morreale, 2021, p. 109). Ospirov notes that "related difficulties in AI training are encountered in styles of music such as atonality, serialism, postserialism, and neotonality" (Ospirov, 2022, p. 37). Therefore, where musical works have higher levels of choice/chance and are "neither exclusive nor contradictory" (Ospirov, 2022, p. 37), machines have difficulty reducing these works into optimal and/or efficient algorithmic models.

A final contributing point concerns Osborne's "anti-aestheticist use of aesthetic materials" (Osborne, 2013, p. 28). Whilst an AI's outputs can be used in ways that humans can transform, can an AI be used in an anti-aestheticist manner if trained on the normed or statistical average? Whilst the algorithm could be innovative, its outputs might be boring. As Laurie Anderson says, "Dangerous art can be made with a pencil" (Anderson, 2016). In other words, challenging work is not a product of a particular software or hardware medium. While most practitioners would prefer to use the best tools in the best and most creative environments, there is also an aspect of making do in creating challenging work.

The value of subverting approaches, therefore, is twofold. The first is that they can generate new materials for artistic works. The second is their necessity in effectively countering the effects of appropriation of musical works – by machines or humans – ideally by avoiding creating musical works and processes which are too easily mimicked and copied.

15.5 Conclusions

For this chapter, I articulated some ideas around the theme of 'you're not supposed to do that', primarily using utility devices, such as a quantizer or vinyl plugin, to produce new outputs not in the devices' original intended purpose.

Practices like dub and Sawari show valuable alternatives to the habitual expressions of creative abuse and failure, which might be unsuited in some practice and teaching scenarios. Likewise, Osborne's "anti-aestheticist use of aesthetic materials" (Osborne, 2013, p. 28) becomes the overarching umbrella term within which these collections of practices can find a stable reference point, which will also correspond with similar activities in other artistic disciplines.

For production tutors, teaching production requires serving multiple needs. Therefore, the practice of going against convention is an attitude that requires cultivation within the education setting, and, as a cultivated practice, it should not be considered optional. In music production, its training and its education, all participants need to be exposed to not just the stable aspects of their craft but also actively encouraged to generate new problems by exploring and subverting devices and processes. Doing so invests a dynamic and challenging approach to their craft, opening up fields of enquiry that may appear closed by the technically focused approach.

Ultimately, 'you're not supposed to do that' becomes a re-articulation of making practices as an essential component for DAW production teaching. This is the quintessential problem for teaching music and production and challenges ideas of reproduction, replay and replication as sufficiently encoded ways of learning. The subversion of conventional tool use is a critical necessity both in the development of new music and production processes and in teaching.

References

Ableton. (no date). Launching Clips, *Ableton* (website), available online from www.ableton.com/en/manual/launching-clips/ [accessed September 2023].

Anderson, L. (2016). Interviewed by Louisiana Channel, *YouTube* (website), , available online from https://youtu.be/dUo-dqMriY8?t=282 [accessed September 2023].

Barker, T. (2007). Error, the Unforeseen, and the Emergent: The Error and Interactive Media Art. *M/C Journal*, Vol. 10, No. 5, doi: https://doi.org/10.5204/mcj.2705.

Baxter, A. (2019). "Clever" Tools and Social Technologies, in Hepworth-Sawyer, R., Hodgson, J. and Marrington, M. (eds.), *Producing Music*, New York: Routledge.

Birch, K. and Muniesa, F. (eds.). (2020). *Assetization: Turning Things into Assets in Technoscientific Capitalism*. Cambridge, MA: The MIT Press.

Boon, H. (2021). Using DAWs as Modelling Tools for Learning Design Sound-Based Applications in Education. *Journal of Music, Technology & Education*, Vol. 13, No. 2–3, pp. 305–322.

Boon, H. (2022). Smear Symphony Demo, *YouTube* (website), available online from https://youtu.be/7foNb9L3pLM [accessed September 2023].

Boon, H. (2023a). 'Live Coding and Music Production as Hybrid Practice', *Organised Sound*, Vol. 28, No. 2, pp. 1–11.

Boon, H. (2023b). InMusic 23- Video Chapter Support – Scale Device, *YouTube* (website), available online from https://youtu.be/FhVvkaQxW0g [accessed September 2023].

Boon, H. (2023c). InMusic 23- Video Chapter Support - Follow Actions, *YouTube* (website), available online from https://youtu.be/U7o5QiBMAM4 [accessed September 2023].

Boon, H. (2023d). InMusic 23- Video Chapter Support - Performance Macro, *YouTube* (website), available online from https://youtu.be/yfCGCt3D0Wc [accessed September 2023].

Boon, H. (2023e). Exploring Art Schema and their Relevance to DAW Teaching and Pedagogy – Examples of Artistic Practice in the Field of Music Production. *Journal of Popular Music Education*, Special Issue: 'Music Technology and Popular Music Education', Vol. 7, No.3, pp. 285–302.

Boon, H. (2024). An Introduction to Neo-Riemannian Operations and Their Application in Music Production, in B. Chapkanov (ed.), *Transformational Analysis in Practice*, Delaware: Vernon Press, pp. 301–320.

Bovell, D. (1979). Interviewed by I. Penman, *NME*, 4 August.

Bovell, D. (2017). Interviewed by C. Melville, *Bass Culture (AHRC Funded Project)*.

Cascone, K. (2000). The Aesthetics of Failure: "Post-Digital" Tendencies in Contemporary Computer Music. *Computer Music Journal*, Vol. 24, No. 4, pp. 12–18.

Chamberlain, A. (2018), Surfing with Sound: An Ethnography of the Art of No-Input Mixing, *Proceedings of ACM Audio Mostly (AM'18)*, Wrexham 12–14 September. https://doi.org/10.1145/3243274.3243289.

Eno Shop. (2010). *Oblique Strategies*, (website), available online from www.enoshop.co.uk/product/oblique-strategies.html [accessed September 2023]

Ewoodzie Jr., J.C. (2017). *Break Beats in the Bronx*. Chapel Hill, NC: University of North Carolina Press.

Gilroy, P. (2017). Interviewed by J. Springer, *Bass Culture (AHRC Funded Project)*.

Hugill, A. (2012). *The Digital Musician*. New York: Routledge.

James, R. (2014). Neoliberal Noise: Attali, Foucault, and the Biopolitics of Uncool. *Cultural Theory and Critique*, Vol. 55, No. 2, pp. 138–158.

Keep, A. (2005). Does Creative Abuse Drive Developments in Music Production? In The Art of Record Production, 17–18 September 2005, University of Westminster, London, UK, (website), available online from www.artofrecordproduction.com/aorpjoom/arp-conferences/arp-archive-conference-papers/17-arp-2005/72-keep-2005 [accessed September 2023].

Kerridge, A. (2016). *Tape's Rolling, Take One! The Recording Life of Adrian Kerridge*. Place of publication unknown: M-Y Books.

Magenta. (no date). Make Music and Art Using Machine Learning, *Magenta* (website), available online from https://magenta.tensorflow.org/ [accessed September 2023].

Mattise, N. (2019). How YACHT Fed Their Old Music to the Machine and Got a Killer New Album. *Ars Technica* (website), available online from https://arstechnica.com/gaming/2019/08/yachts-chain-tripping-is-a-new-landmark-for-ai-music-an-album-that-doesnt-suck/ [accessed September 2023].

McNamee, D. (2009), 'Hey, what's that sound: Oblique strategies', *The Guardian*, 7 September. www.theguardian.com/music/2009/sep/07/ oblique-strategies. Accessed 04 August 2024.

Metheny, P. (1985). Interviewed by N. Webb, *Guitar Player* (website), available online from www.joness.com/gr300/metheny.htm [accessed September 2023].

Morreale, F. (2021). Where Does the Buck Stop? Ethical and Political Issues with AI in Music Creation. *Transactions of the International Society for Music Information Retrieval*, Vol. 4, No. 1, pp. 105–113, https://doi.org/10.5334/tismir.86.

Moore, A. (2019). Tracking With Processing and Coloring as You Go, in Hepworth-Sawyer, R., Hodgson, J. and Marrington, M. (eds.), *Producing Music*, New York: Routledge.

Moorefield, V. (2010). *The Producer As Composer*. MA: MIT Press.

Numan, G. (2020). *(R)evolution: The Autobiography*. London: Little, Brown Book Group.

Osborne, P. (2013). *Anywhere Or Not At All: Philosophy of Contemporary Art*. Brooklyn, New York: Verso Books.

Ospirov, A. (2022). The Student – Shortcuts Guide To Music Theory, in Clancy, M. (ed.), Artificial Intelligence and Music Ecosystem, Abingdon, Oxon: Routledge.

Phillips, A. and Taylor, B. (1980). Sex and Skill: Notes towards a Feminist Economics. *Feminist Review*, Vol. 6, pp.79–88.

Runway ML. (2023). Advancing Creativity with Artificial Intelligence, *Runway* (website), available online from https://runwayml.com/ [accessed September 2023].

Takemitsu, T. (1995). *Confronting Silence: Selected Writings*. MD: Scarecrow Press.

Théberge, P. (2001). 'Plugged In': Technology and Popular Music, in Frith, S., Straw, W. and Street, J. (eds.), *The Cambridge Companion to Pop and Rock*, Cambridge: Cambridge University Press, pp. 1–25.

Trio Sawari. (2017). 27 Questions For A Start . . . And Some Possible Answers to Begin With, in Cox, C. and Warner, D. (eds), *Audio Culture: Readings in Modern Music*, London: Continuum, pp. 407–16.

Williams, S. (2012). Tubby's Dub Style: The Live Art of Record Production, in Frith, S. and Zagorski-Thomas, S. (eds.), *The Art of Record Production: An Introductory Reader for a New Academic Field*, Aldershot: Ashgate, pp. 235–246.

Zagorski-Thomas, S. (2014). *The Musicology of Record Production*. Cambridge: Cambridge University Press.

16

ARTISTIC INTUITION AND ALGORITHMIC PREDICTION IN MUSIC PRODUCTION

Mads Walther-Hansen

16.1 Introduction

AI-based tools for music production (e.g., automated effect plugins for mixing, chord generators, and sound synthesis tools) are rapidly developing and have the potential to influence future music production fundamentally. A survey (n1500) from 2023 among readers of Bedroom Producers Blog (2023) showed that 36,8 percent of the surveyed producers used AI tools as part of their setup, and 30 percent intended to try them in the future. Sometimes the tools are used as assistive devices that music producers bring into their existing workflow to make suggestions and aid decisions; other times, they function as fully automated tools that respond autonomously to set targets or aims (Moffat and Sandler, 2019).

While digital tools for recording and mixing have long enabled music producers to automate different production tasks, for instance, by using autotune, new AI tools are delegated decision tasks that mimic the result of human reflection – including decisions that can be characterized as artistic (e.g., automatic adjustment of tonal balance or automatic generation of new sounds.). This delegation may reduce the producer's cognitive workload, speed up the production process, and generate new ideas (see Deruty et al., 2022). Still, AI's ability to innovate (when prompted to act autonomously) is limited to the patterns it has learned from the data it was trained on (Slota et al., 2020). It can learn from, mimic, and merge existing styles and patterns. However, it is questionable whether AI has the same capacity for intuitive leaps as humans, that often stumble upon innovative ideas through serendipity or gut feeling (Johanssen and Wang, 2021).

The history of music production is rich with examples of new innovative sounds that emerged from intuitive leaps, technical malfunctioning, accidental misuse of technology, or other incidental events in the recording studio.

For instance:

- In 1951, guitarist Willie Kizarts plugged in his defective guitar amplifier during a session with Jackie Brenston and his Delta Cats in Sun Studios. After failed attempts to fix the cone with a piece of paper, producer Sam Philips realizes that what he is hearing is a valuable and original sound – the sound of the distorted guitar.

DOI: 10.4324/9781003396710-16

- In an interview from 2001, producer Marius De Vries recalls a studio session with Björk, where she responded to a sound by saying: "Hold a pineapple in your hand and look at the fluffy bit at the top, well it needs to sound more like that" (De Vries in Flint, 2001).

Björk's deliberate metaphorical play with words that triggered Marius De Vries' imagination, and the malfunctioning amplifier that inspired Sam Philips led to innovative sounds. These sounds emerged from extraordinary situations where the producers felt inspired to do something different – something *they were not supposed to do* and something that they could not have foreseen they would do before the recording session.

Contrary to this, current AI-assisted music production technologies appear to tell you what you *are supposed to do*. They are technologies based on sophisticated statistical techniques that use historical data to enable a prediction (see Agrawal et al., 2022) – for instance, to predict the music producer's preference (and eventually the music consumers' preferences) for specific effect settings.

Some researchers have expressed concern about the consequences of AI in the arts, for instance, that it may create more homogenization and standardization (Esling and Devis, 2020; Brook, 2023). Others explore the possible benefits of AI that may challenge and improve current practices by providing new insights and optimizing processes in the recording studio (e.g., De Man et al., 2020).

This paper does not attempt to provide an exhaustive analysis of AI's effect on music production practices. Instead, the purpose is to reflect on the future conditions for artistic innovation in sound (i.e., the introduction of new and original ideas and sounds) in music production in the age of AI. To do this, first, the paper discusses how AI may change music production as an art form, and it is suggested that music producers must learn to navigate the gap between human intuition and AI predictions. Second, building on the extended-mind paradigm, three metacognitive strategies are outlined – *internal*, *offloading*, and *outsourcing* – to illustrate how AI-based tools may shape reasoning and action in the recording studio. Finally, the chapter concludes with reflections on possible future consequences of more AI-based decisions in music production.

16.2 Music Production As Artistic Work

Researchers have referred to music production as an art form for over four decades. In his seminal article from 1979, "From Craft to Art – The Case of Sound Mixers and Popular Music," the sociologist Edward R. Kealy argued that the work of music producers (Kealy called them sound mixers) involved decisions that should be considered artistic rather than merely technical. According to Kealy, the role of music producers changed gradually with the creative studio practices surrounding the rock n' roll and rock eras in the 1950s and 1960s: "recording artists began annexing the craft of sound mixing to their art, while some sound mixers attempted to slough off their designation as "technicians" and to establish a new collaborative role as "artist-mixers" (Kealy, 1979, p. 4). Consequently, "the degree to which sound mixers have taken part in aesthetic decision making has increased" (Kealy, 1979, p. 7).

Considering Kealy's depiction of how practices in the recording studio became more artistic, how should we value the work of future music producers that (possibly) will rely more on algorithmic decision-making? Will AI develop music production in new unforeseen directions, or is the art of music production about to innovate itself to death?

It is tempting to conclude that music production will gradually return to a craft-based or technical discipline. However, this conclusion requires that we accept the distinction between artistic work and craft-based practices that Kealy derives from Howard Becker (1978). According to Becker (1978), artistic work is informed by an aesthetic ideology that distinguishes it from mere craft. Craftworkers produce applicable creations. Something that serves a practical purpose. Thus, creating craft often involves a high degree of standardized processes. On the contrary, artistic practices are typically related to actions that lead to creations with unique qualities and aesthetic value.

That the art-craft distinction is problematic, fluid, and sometimes deceptive has long been evident in the philosophy of the arts (e.g., see Collingwood, 1938). Still, it has remained a convenient way to categorize the value of different products and creative processes. The distinction is no less problematic to account for actions and reasoning in the recording studio, where aesthetic reasoning (e.g., attention to sensory opportunities in sound) and more functional reasoning (e.g., attention to technical matters, market demands, and so on) often coincide.

AI has introduced new perspectives to this debate. Agrawal et al. (2022) define AI as a significant advance in statistical techniques that enables a prediction. It is inclined to treat sensory inputs as media for strategic decisions (e.g., analyzing a situation, identifying challenges, and developing a plan to achieve specific goals). The predictions delivered by AI are not absolute certainties but rather probabilities or likelihoods based on the training data. However, as these probabilities improve, predictions may be increasingly valuable to make informed decisions in the recording studio, and AI may make decisions without human interference.

Using its prediction power, AI can make (or contribute to) products that resemble human artistic work. AI algorithms can predict how different EQ and compressor settings will affect the final mix, predict optimal settings for balance and clarity, make personalized predictions about a particular music producer's preference, and more. This allows for music productions made entirely by AI that – in their physical manifestation – are indistinguishable from music productions made by humans (good and bad ones).

Whether AI art is proper art or not has been discussed elsewhere (e.g., Still and d'Inverno, 2019; Miller, 2019), and it is not the goal of this article to further muddle this discussion. Suffice it to say that AI cannot reproduce the reasoning process that leads to human art. It lacks embodied interoceptive processes underlying human emotions (Novelli and Proksch, 2022), and it is never caught up in the moment or inspired by something that lies outside the training set (see Manovich, 2018). Thus, it is fair, at least, to characterize AI art as a separate kind of art that does not rely on human intuition, imagination, and inspiration (although it may attempt to mimic these aspects of human reasoning).

Here, it is discussed how AI may inform and challenge the artistic process in the recording studio as an extension of the music producer's cognitive environment. To do this, it is necessary, first, to explain a few key differences between AI predictions and human reasoning.

16.3 Intuitive and Predictive Decisions

In 1959, two significant but unrelated events occurred in the fields of computer science and music production: At IBM in New York, the American computer scientist Arthur Samuel (1959) coined the term machine learning when he described the process used in his checkers-playing program. Samuel aimed to make computers adapt and improve their performance over time based on the structures and patterns they identify in a dataset. In this way:

A computer can be programmed so that it will learn to play a better game of checkers than can be played by the person who wrote the program . . . when given only the rules of the game, a sense of direction, and a redundant and incomplete list of parameters which are thought to have something to do with the game, but whose correct signs and relative weights are unknown and unspecified.

(Samuel, 1959, p. 535)

In Detroit, the young songwriter and record producer Berry Gordy Jr. bought a property, named it Hitsville USA, and formed the Tamla (later Motown) label. Cogan and Clark (2003) describe the session leading to one of the first Tamla hit recordings published the same year, Barrett Strong's "Money (That's What I Want)" (1959), as a series of coincidental events. While Gordy was jamming on the piano, Strong, working in another room, heard the music and came to sing along. According to Strong (quoted in Cogan and Clark 2003), two young men he did not know came by the studio and asked if they could join the session on bass and guitar. Also present were the studio drummer Benny Benjamin and the songwriter Brian Holland, who picked up a tambourine and jammed along with the rest of the group. Gordy finally decided to capture the performance live on a two-track tape recorder. While the song itself might not be considered a quintessential representation of what should be known as the Motown Sound, the production session played a significant role in shaping the label's early artistic direction, with, for instance, catchy melodic hooks, a strong rhythm section, and frequent use of tambourine.

In an interview from 2014, Gordy described how his decisions were often based on intuition: "I just went by my intuition, and I have always been a risk-taker – and if I believed in something, I would stick with it until it worked" (Gordy, 2014). This description aligns with other notable accounts of the artistic process in the recording studio. Zak (2022), for instance, describes music production as "an array of decisions informed by artistic intuition as well as experienced technique," and Burgess (2013) highlights "the artistic blending of intuition, rationality, and experience" (Burgess, 2013, p. 264) as the essence of music production.

While intuition is only one among many aspects involved in decisions in the recording studio (music producers typically have a range of analytical, technical, and communicative skills acquired from training and experience), it is of central importance to innovation in sound (see also Weintraub, 1998). Human intuition has been characterized as "one of the hallmarks of how human beings think and behave" (Sadler-Smith, 2008) – a non-conscious aspect of cognition that has an invaluable role in the arts (Reid, 1981). An intuition-based decision results from an instinctive understanding of something that does not rely on intentional reasoning. The guiding principles for decisions are gut feelings, emotions, past experiences, and other subconscious processes. Relying on your intuition is – as Theresa Jane Hardman (2021) phrases it – a way to cultivate an "attitude of openness to uncertainty and relinquishing control of outcomes, as well as thinking in different ways, [that] leads to [a] revelation of new possibilities" (Hardman, 2021, p. 4).

The machine-based reasoning that Samuel worked on and the intuition that guided (and still guides) artistic processes in many recording studios can be characterized as two fundamentally different approaches to decision-making. AI is usually associated with computer systems that can perform tasks that we – until recently – assumed required human thinking (many past AI technologies that have become commonplace, such as Samuel's game-playing machine, are no longer characterized as AI). It analyzes data and uses statistical models and machine learning to make informed predictions that help maintain a stable, reliable, and (sometimes) explainable decision-making process (Miller, 2019).

Historically music producers would often find themselves in situations where they have little but their intuition to rely on. It would have been difficult (if not impossible) for Sam Philips to explicitly justify why the distorted guitar should be kept in the recording of "Rocket 88" or for Berry Gordy to justify why the sound in "Money (That's What I Want)" would work with listeners.

They – allegedly – had the courage (and the power qua their position) to go with what they felt was right.

Today the situation is different. Contrary to pre-AI studio technologies, many studio plugins today not only tap into decision-making as interactive tools that inform human reasoning (Walther-Hansen and Eskildsen, 2024). They can execute decisions and actions independently. Following this development, why should music producers today be concerned with intuition when they can access AI-powered predictions? In the following discussion, it is argued that the next generation of technologies will put increasing demands on the music producer's metacognitive decisions (see Flavell, 1979) to balance the gap between relying on intuition or leaving decision and task processing to AI-based predictions.

16.4 Ai and The Extended-Mind Paradigm – From Offloading To Outsourcing

In 1998, Clark and Chalmers proposed the idea of the extended mind, described as a form of active externalism, where components in the environment take an active role in driving cognitive processing. They argued that "the human organism is linked with an external entity in a two-way interaction, creating a coupled system that can be seen as a cognitive system in its own right" (Clark and Chalmers, 1998, p. 8). If we believe with Clark and Chalmers that our cognitive system consists of both mind-internal and mind-external components, then how are reasoning and action in music production changing with the introduction of still more sophisticated AI tools?

We listen, think, and act using our bodies and our environment. Listening and decision-making in the recording studio is a process that happens over time, where the music producer experiments with different effect settings, dynamic balance, and so on. The producer is exposed to auditory stimuli and is surrounded by technology that affords (cf. Gibson, 1979; Clarke, 2005) certain kinds of listening and responses. These stimuli do not *prescribe* how to listen and respond, but they *limit* the range of possible decisions and actions.

But perceptual information not only guides experience and action. The environment is used to store and aid cognitive operation. Music production is a complex task, and there are limits to our mind-internal cognitive capabilities (cognition is time-pressured and has limited processing power and storage capabilities). For this reason, mental computation is often *offloaded* (see Risko and Gilbert, 2016) onto the environment. It is often faster and puts less constraint on cognition to turn a set of knobs and listen to the effect than to mentally rehearse a vast number of settings that might lead to the desired sound (see also Kirsh and Maglio's [1994] study of how Tetris players use rotations on the screen to offload cognition, because it is faster and less cognitively demanding to rotate the pieces to find a fit than it is to rehearse the rotations mentally).

In cognitive sciences, offloading is usually understood as a process where the environment is used to aid and improve cognitive processing (e.g., memory and reasoning) when performing a task (Walther-Hansen and Grimshaw, 2016; see also Risko and Gilbert, 2016, for an overview). Offloading entails that the extended cognitive system – consisting of mind-internal and mind-external components – functions as an integrated system (Ienca, 2018). It is a metacognitive strategy (see Lefford and Thompson, 2018, for a broader discussion of metacognitive strategies in the recording studio) to optimize cognitive competencies. For instance, moving faders and knobs on the mixing desk is a means to reason faster and more efficiently to decide on the best setting. One may also store settings that can be retrieved later to aid memory, for instance, to postpone further processing until one has the internal cognitive capacity to deal with it. Through continuous interaction with studio technology, internal cognitive demand is reduced in various ways, and overall cognitive performance is improved.

Using AI-assisted technology to make decisions in an artistic process seems to be another – and perhaps more significant – form of offloading than, for instance, using technology as mind-external memory and an external source to aid reasoning. Using AI, the music producer may not only reduce cognitive workload; it is possible to (nearly) eliminate it. Thus, the term *offloading* does not embrace the complete understanding of the dynamic relationship between AI decisions and the user's mind-internal decisions.

Many AI-based music production effects function as *independent systems* that not only analyze and suggest ways to solve issues; they perform the task (see Moffat and Sandler, 2019). These *independent systems* are not components of our extended cognitive system per se. When a music producer uses an *extended strategy* (Risko and Gilbert, 2016) and assigns these systems a task, they work autonomously until the task is finished. In such cases, using AI may be no different from handing over a task to another human agent (e.g., a studio assistant).

To explain this situation, it is useful to distinguish between internal strategies and two forms of extended strategies: 1) *offloading* (i.e., a cognitive process where components in the environment store or process a pool of cognitive information that the user will interact with, retrieve when needed, or otherwise use to complete a task) and 2) *outsourcing* (i.e., a process where a task – and the required decisions to complete the task – is entirely handed over to full automatic AI technology, see Figure 16.1).

AI is used as a form of *cognitive outsourcing* when we get it to do something on our behalf (see Danaher, 2018). It includes situations where AI tools complete the entire mix or perform sub-tasks, for instance, when the music producer decides to outsource the tonal balancing of a track to an independent AI system.

It should be acknowledged that the lines are blurred between the three forms of metacognitive strategies – internal strategies (e.g., relying on intuition or other forms of mind-internal reasoning over AI prediction), offload strategies (prompting AI to suggest settings the music producer can reject, accept, or modify) and outsource strategies (prompting AI to decide settings and complete the task) – and the music producer may alternate between them during a production session. Here the categories are presented as a methodological tool to understand decision-making in the recording studio.

Lefford and Thompson (2018) have previously explored music producers' use of metacognitive strategies in the recording studio in the context of a macrocognitive setting – that is, the natural

Internal strategies	Extended strategies	
	Offload strategies	Outsource strategies
 • Mind-internal components drives information and task processing • No reduction of internal mental processing • Decision-making only involve mind-internal processes	 • Mind-external components are used to aid information and task processing • Reduces the burden on internal mental processing • Decision-making involve mind-internal and mind-external processes	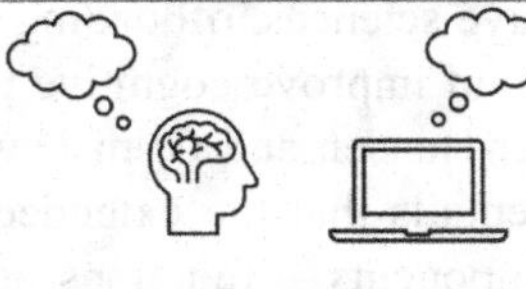 • Mind-external components take over information and task processing. • Removes the burden from internal mental processing • Decision-making only involve mind-external processes

FIGURE 16.1   Internal and extended strategies for decision-making.

environment of the recording studio and the related conditions, for instance, resource constraints, the uncertainty of the creative process, and the pressure to perform well and deliver high-quality recordings and mixes. Observing different forms of decisions, the authors show how music producers use different forms of metacognitive skills to regulate and manage cognitive demands.

I suggest here that studio environments with AI tools complicate the basis for decisions even further, requiring music producers to develop new forms of metacognitive skills. On the one hand, the decision to use outsource strategies may be based on *internal demands*, such as cognitive overload or low confidence in a particular task (see Risko and Gilbert, 2016). On the other hand, the decision to use a particular AI system to complete a task may depend on the user's confidence in the system (Weis and Wiese, 2022). Thus, deciding whether to use outsource strategies over internal or offload strategies requires the music producer to weigh the relation between self-confidence in the task and trust in the AI system. For instance, if a producer has little self-confidence in a task and feels confident that the technology can perform better, it may call for outsource strategies. It is a tool to reduce the fear of failure.

Extended strategies may also have long-term consequences for the music producer's metacognitive behavior (see Grinschgl and Neubauer, 2022). While AI provides solutions that producers (and consumers) are likely to accept as good, it also teaches producers what they *should* accept as reasonable. Thus, after repeatedly choosing to outsource decisions to AI, music producers may rely more on AI predictions than internal strategies.

16.5 Coda

AI-based music production technologies are still emerging, and only recently have we started to discuss their possible effects on decision-making in the recording studio and, more broadly, how they may change the art of music production. This paper focused on the tension between human intuition and AI predictions, and it was argued that music producers should learn to navigate between different forms of decision-making. This requires new forms of metacognitive skills.

We may assume that music producers will rely more on outsourcing strategies when AI starts to predict music producers' (and music consumers') preferences more accurately. Photographers and filmmakers already rely heavily on AI technologies such as face recognition, auto color adjustment, auto-zoom, and so on. This allows amateur photographers armed with a smartphone to outsource editing of all these parameters. Compared to this, the audio industry has been much slower to develop and adapt new technologies (Moffat and Sandler, 2019).

Suppose decisions in the recording studio were only rational and based on market demands, current trends, and other music-external factors, then human intuition – subject to biases and errors – would be lacking in several areas. Relying solely on internal strategies would lead to suboptimal decisions and judgments. However, decisions in the recording studio are not only rational and sometimes not at all. Decisions are based on a complex (and sometimes contradictory) combination of parameters (see Lefford and Thompson, 2018), including artistic visions, technological skills, expected audience preferences, time constraints, available resources, and much more – a process that often leads to unintended outcomes.

Intuitive and unforeseen leaps of reasoning have always contributed to new ideas, genres, and techniques in the recording studios (see Cascone, 2000). Sometimes detours from conventional practices are caused by technical malfunctions or other accidental events, as in the case of the recording of the (allegedly) first distorted guitar. In other cases, collaborators encourage deviations from standard routines, such as when Björk urged Marius de Vries to make a sound like the top of a pineapple (see Section 17.1). Such processes are often challenging, cumbersome, and time-consuming. Björk (deliberately, we may assume) complicates the sound editing process to exercise the producer's imagination and allow new cognitive images of sound to form (see Walther-Hansen, 2024). Contrary to most AI tools for music production that producers may use to make the task easier and more efficient (i.e., to either offload or outsource cognitive processes), she makes the producer's cognitive burden more extensive. From the producer's perspective, it is a form of *cognitive onloading*.

Stories like this may tell us that artists will find ways to re-introduce the tension and inspiration of artistic work once effectiveness and automation take over. We may see new movements in music production that focuses on slowness, ineffectiveness, and disordered thinking to counteract the new tendencies (not unlike how analog technologies were re-introduced in recording studios in the 1990s as a reaction to digitization – see, e.g., Bennett, 2012; Walther-Hansen, 2020). More research is needed in the future to understand how music producers with direct access to AI-powered prediction balance internal and external strategies in decision-making.

References

Agrawal, A. Gans, J. and Goldfarb, A. (2022). *Power and Prediction: The Disruptive Economics of Artificial Intelligence*. Boston: Harvard Business Review Press.

Becker, H. S. (1978). Arts and Crafts. *American Journal of Sociology*, Vol. 83, No. 4, pp. 862–889.

Bedroom Producers Blog. (2023). https://bedroomproducersblog.com/2023/05/30/ai-music-survey/#:~:text=Currently%2C%2036.8%25%20of%20producers%20use,are%20disappointed%20by%20their%20quality (accessed August 2023).

Bennett, S. (2012). Endless Analogue: Situating Vintage Technologies in the Contemporary Recording & Production Workplace. *Journal on the Art of Record Production*, Vol. 7. www.arpjournal.com/asarpwp/endless-analogue-situating-vintage-technologies-in-the-contemporary-recording-production-workplace/

Brook, T. (2023). Music, Art, Machine Learning and Standardization. *Leonardo*, Vol. 56, No. 1, pp. 81–86.

Burgess, R. J. (2013). *The Art of Music Production – The Theory and Practice*. New York: Oxford University Press.

Cascone, K. (2000). The Aesthetics of Failure: "Post-Digital" Tendencies in Contemporary Computer Music. *Computer Music Journal*, Vol. 24, No. 4, pp. 12–18.

Clark, A. and Chalmers, D. J. (1998). The Extended Mind. *Analysis*, Vol. 58, No. 1, pp. 7–19.

Clarke, E. F. (2005). *Ways of Listening: An Ecological Approach to the Perception of Musical Meaning*. New York: Oxford University Press.

Cogan, J. and Clark, W. (2003). *Temples of Sound: Inside the Great Recording Studios*. San Francisco: Chronicle Books.

Collingwood, R. G. (1938). *The Principles of Art*. London: Oxford University Press.

Danaher, J. (2018). Towards an Ethics of AI Assistants: An Initial Framework. *Philosophy and Technology*, Vol. 31, No. 4, pp. 629–653.

De Man, B., Reiss, J. and Stables, R. (2020). *Intelligent Music Production*. New York: Routledge.

Deruty, E., Grachte, M., Lattner, S., Nistal, J. and Aouameur, C. (2022). On the Development and Practice of AI Technology for Contemporary Popular Music Production. *Transactions of the International Society for Music Information Retrieval*, Vol. 5, No. 1, pp. 34–49.

Esling, P. and Devis, N. (2020). Creativity in the Era of Artificial Intelligence. *ArXiv.* /abs/2008.05959.

Flavell, J. H. (1979). Metacognition and Cognitive Monitoring: A New Area of Cognitive-Developmental Inquiry. *The American Psychologist*, Vol. 34, No. 19, pp. 906–911.

Flint, T. (2001). Musical Differences – Marius De Vries. *Sound on Sound Magazine*, November.

Gibson, J. J. (1979). *The Ecological Approach to Visual Perception*. Boston: Houghton Mifflin Company.

Gordy, B. (2014). Interview with Berry Gordy. On "Windy City Live" (unknown date). Chicago. www.youtube.com/watch?v=gdqdBoSqdK0 (Accessed August 2023).

Grinschgl, S. and Neubauer, A. C. (2022). Supporting Cognition with Modern Technology: Distributed Cognition Today and in an AI-Enhanced Future. *Frontiers in Artificial Intelligence*, Vol. 5, Article 908261.

Hardman, T. J. (2021). Understanding Creative Intuition. *Journal of Creativity*, Vol. 31, Article 100006.

Ienca, M. (2018). An Evolutionary Argument for the Extended Mind Hypothesis. *Orbis Idearum*, Vol. 6, No. 1, pp. 39–61.

Johanssen, J. and Wang, X. (2021). Artificial Intuition in Tech Journalism on AI: Imagining the Human Subject. *Human-Machine Communication*, Vol. 2, pp. 173–190.

Kealy, E. R. (1979). From Craft to Art: The Case of Sound Mixers and Popular Music. *Work and Occupations*, Vol. 6, No. 3, pp. 3–29.

Kirsh, D. and Maglio, P. (1994). On Distinguishing Epistemic from Pragmatic Action. *Cognitive Science*, Vol. 18, No. 4, pp. 513–54.

Lefford, M. N. and Thompson, P. (2018). Naturalistic Artistic Decision-Making and Metacognition in the Music Studio. *Cognition, Technology & Work*, Vol. 20, No. 4, pp. 543–554.

Manovich, L. (2018). *AI Aesthetics*. Moscow: Strelka Press.

Moffat, D. and Sandler, M. B. (2019). Approaches in Intelligent Music Production. *Arts*, Vol. 8, No. 125, pp. 1–14.

Miller, A. I. (2019). *The Artist in the Machine: The World of AI-Powered Creativity*. Cambridge: MIT Press.

Novelli, N. and Proksch, S. (2022). Am I (Deep) Blue? Music-Making AI and Emotional Awareness. *Frontiers in Neurorobotics*, Vol. 16, Article 897110.

Reid, L. A. (1981). Intuition and Art. *The Journal of Aesthetic Education*, Vol. 15, No. 3, pp. 27–38.

Risko, E. F. and Gilbert, S. J. (2016). Cognitive Offloading. *Trends in Cognitive Sciences*, Vol. 20, No. 9, pp. 676–688.

Sadler-Smith, E. (2008). *Inside Intuition*. London: Routledge.

Samuel, A. L. (1959). Some Studies in Machine Learning using the Game of Checkers. *IBM Journal of Research and Development*, Vol. 3, No. 3, pp. 210–229.

Slota, S. C., Fleischmann, K. R., Greenberg, S., Verma, N., Cummings, B., Li, L. and Shenefiel, C. (2020). Good Systems, Bad Data? Interpretations of AI Hype and Failures. *Proceedings of the Association for Information Science and Technology*, Vol. 57, No. 1, p. e275.

Still, A. and d'Inverno, M. (2019). Can Machines Be Artists? A Deweyan Response in Theory and Practice. *Arts*, Vol. 8, No.1, Article 36.

Walther-Hansen, M. (2020). *Making Sense of Recordings: How Cognitive Processing of Recorded Sound Works*. New York: Oxford University Press.

Walther-Hansen, M. (2024). Enacting the Environment Through Sound – Reflections on the Use of Cognitive Metaphors in Sound Interaction Design. In Filimowicz, M. (ed.) *Routledge Handbook of Sound Design*, pp 37–48. New York: Routledge

Walther-Hansen, M. and Eskildsen, A. (2024). Forceful Action and Interaction in Non-Haptic Music Interfaces. In Gullö, J.-O., Hepworth-Sawyer, R., Paterson, J., Toulson, R., and Marrington, M. (eds.) *Innovation in Music – Music Production: International Perspectives*, pp. 256–266. London: Routledge.

Walther-Hansen, M. and Grimshaw, M. (2016). Being in a Virtual World: Presence, Environment, Salience, Sound. *Proceedings of the 11th Audio Mostly Conference*. ACM Digital Library, pp. 77–84.

Weis, P. P. and Wiese, E. (2022). Know Your Cognitive Environment! Mental Models as Crucial Determinant of Offloading Preferences. *Human Factors*, Vol. 64, No. 3, pp. 499–513.

Weintraub, S. (1998). *The Hidden Intelligence: Innovation Through Intuition*. Boston: Butterworth-Heinemann.

Zak, A. (2022). The Art of Record Production (Editorial). *Journal of the Art of Record Production*, Vol. 2. www.arpjournal.com/asarpwp/the-art-of-record-production/

17

A RADIOLOGICAL ADVENTURE

The Sonification of the Apocalypse

Charles Norton, Daniel Pratt, and Justin Paterson

17.1   Introduction

This chapter encapsulates a research journey that delves into the intersection of technology and culture, as viewed through the lens of the theme: 'You are not supposed to do that', which has been interpreted as an invitation to explore the fringes of possibility. Consequently, the central question at the heart of this exploration is, 'Can a Geiger-Müller Counter be connected to a synthesiser?' The brief response is an affirmative 'yes'. While the technical and programmatic intricacies of 'how' this is achieved will be detailed later in the exegesis, a more profound initial enquiry arises: 'Why?' This question is examined through a chronology constructed from various sources. We seek to uncover the motivations, biases, and contradictions inherent in society's relationship with nuclear technology. This work may be viewed through the lens of subtle activism rather than solely as an experimental musicological pursuit.

17.2   Methodology

An experimental, ethnographic, and auto-ethnographic narrative timeline is presented. Taking a reflective approach (Gonot-Schoupinsky, 2022) provides a route to question and explore researcher bias in this cultural investigation (Adams et al., 2022). This process involved mind mapping, research, and reflection cycles, which revealed a set of fictional and historical influences from childhood to the present. Each influence has been investigated, revisited, and annotated. The data gathered collectively forms a narrative that establishes the creative context and provides perspective on the decision-making.

Secondly, practice-based research is framed within practical musicology as described by Zagorski-Thomas (2022) in which the allographic (a set of instructions, a reproducible act) and autographic (a unique artefact) outputs are detailed and deconstructed.

DOI: 10.4324/9781003396710-17

17.3 Background and Related Work

17.3.1 Geiger Müller Counter

The Geiger-Müller Counter (GMC) is a scientific instrument that has become a central component within a creative ecosystem. Hans Geiger was the first to detect particles successfully, and his student Walther Müller refined the detector (Shampo et al., 2011). Designed to measure radiation levels, it indicates when isotopes pass through the gas in a cathode tube, creating a voltage change (Knoll, 2000). This event is announced by a beep or click from the device. This allows radiation levels to be monitored without looking at the display constantly: a direct sonification. The device keeps count of the events and provides a statistical measurement of the data. Using MaxVCS to map the GMC data collected in different situations and using a prototype hardware interface, a realm of creative inspiration has been drawn. Ahmed (2007) presents a foundational text regarding radiation and detection: an unlikely source of inspiration. However, this reference provides an understanding of how the GMC operates, what it measures, and its inherent limitations. Intriguingly, sections of Ahmed's book delve into technological intricacies that resonate with audio synthesis and recording; Nyquist, sample rate, aliasing (Roads, 1996), and other terminology find common ground. Ahmed characterises some aspects of the GMC's dosimeter as a 'Time to Voltage Converter', a concept that will become relevant shortly.

17.3.2 MaxVCS

This chapter builds upon the research presented in '*Innovation In Music: Technology and Creativity*' (Gullo & Hepworth-Sawyer et al., 2024), which explored parameter-mapping strategies for electronic musicians. Motivated by the constraints of the MIDI 1.0 and the slow adoption of the improved MIDI 2.0 protocols: An innovative mapping system has been designed, utilising OSC to enable a bidirectional connection with minimal latency and low jitter between Max8 (Cycling 74, 2020) and Kyma (Symbolic Sound, 2023). The MaxVCS framework addresses existing challenges and amplifies the connectivity between the two platforms. As we consider prospective directions and further development, streamlining and pinpointing the MaxVCS software design's primary goal has brought both programmatic dilemmas and philosophical inquiries to the fore. 'An Ode to Prototyping' (2023) presents an introduction to the context.

The birth of this system was fuelled by a desire for enhanced connectivity and complexity, aiming to aid a prototyping process encompassing physical, programmatic, and haptic elements. The transition from 'alpha' to 'beta' versions of the software necessitates a simple, hands-free

FIGURE 17.1 An ode to prototyping.

voltage source for testing, ideally non-repetitive, to facilitate concentrated mapping without needing continual intervention or succumbing to monotony. A radiation counter (details to follow) was procured featuring an exposed circuit board, presenting an irresistible opportunity for voltage probing and hi-jacking of the Geiger 'ping' for experimental purposes. MaxVCS transformed from proof of concept 'alpha' to a viable powerful tool integral to this creative process.

Whilst grappling with the electrical and programmatic obstacles, it provided time for introspection regarding the motivations fuelling this venture. Initially perceived as a trivial diversion, a deeper analysis uncovered a complex tapestry of personal experiences, cultural impacts, and historical occurrences illuminating the varied influences and inherent biases steering this creative expedition.

17.4 Sonification Of The Apocalypse

This title (which will be discussed in the exegesis) refers to a holistic body of work which comprises three elements.

1. Textual: The text blends personal experience, fiction, and historical events to unpick the layers of influence that culminated in purchasing the GMC and considering its creative potential.
2. Technological: Allographic techniques.
3. Audio and Visual: Autographic artefacts.

The second two elements are blurred in distinction to a certain extent due to the creative feedback loop.

17.4.1 Prologue

My nine-year-old brain was startled as the proclamation 'American President Ronald Reagan and Russian President Mikhail Gorbachev were at Loggerheads' blared from the television (Lee, 1992). Living in a village called 'Loggerheads' in the middle of the Midlands, far away from anything, the concept that Reagan and Gorbachev were in the village was mind-boggling. Obviously (to the rest of the planet), the superpowers were not discussing nuclear disarmament in our local pub. Nevertheless, the surreal sensation of hearing our village's name on the news left an indelible mark. Curiosity piqued, I explored the etymology of 'Loggerheads', discovering its origin in the aftermath of the Battle of Blore Heath, a pivotal event during the 'War of the Roses' on September 23, 1459. This historical connection, combined with the realisation that the world-altering conversations weren't taking place on our doorstep, ignited an understanding of my role as a citizen – not just within the United Kingdom but as a member of a broader civilisation replete with its historical legacies, tensions, and disputes. This memory is the genesis of this radiological narrative.

17.4.2 Raymond Briggs

Our household Christmas traditions included watching classics such as: *The Sound of Music* (1965), *A Grand Day Out* (1994), and *The Snowman* (1982). The latter was a particular childhood favourite, so when the TV announcer uttered the name 'Raymond Briggs', I watched with festive glee. What followed was not the familiar melancholic yet heart-warming production but the existentially chilling *When the Wind Blows* (1986) (WTWB). Adapted from Briggs's novels, these classic animations demark a threshold in my emerging global perspective. The contrast between the expected merry spectacle and the experienced narrative was profound. Set in the same rolling hills as *The Snowman*, WTWB follows Jim and Hilda, an average couple living in the 'home counties' during a nuclear strike. Several sections struck resonance while reviewing the animation. Firstly, stockpiling; the couple struggles to gather basic supplies. These scenes echo recent global supply chain issues caused by digital disruption (ncsc.gov.uk, 2018), biological (BBC TV News, 2020), and political (Walker, 2023). Secondly, Peggy discusses the radiation: 'If you can't see it and can't feel it, it can't be doing you any harm, can it!' This line haunts me, and I internally scream, 'But we can measure it!' This was the genesis of my curiosity regarding radiation measurement. A lasting moment was the spectacle of the Russian figure presented in a Cold War aggressive stance, which seemed laughable in the context of the disarmament discussions of the time yet entirely sobering now. A linguistic observation regards a word Jim reads from the official literature: 'Megadeath'. My only previous association was with the band 'Megadeth'; I had no idea regarding the etymology. After Dave Mustaine was fired from Metallica in 1983, he needed to write some lyrics onto something, picking up a handbill produced by Senator Cranston discussing nuclear weapons, including the phrase: 'The arsenal of megadeath can't be rid' (Wiederhorn, 2022). I find it fascinating how such words of horror permeate into popular culture and take on new meanings in new contexts. 'Megadeath', first recorded in 1953, means the death of a million people (merriam-webster.com, 2023). Another definition is even more chilling, vocabulary.com (2023) presents: 'They calibrate the effects of atom bombs in megadeaths.' If that is not macabre enough, then we can rely on etymonline.com (2023) to horrify: 'The resulting one million dead bodies is a megacorpse.'

17.4.3 Narrative Impact

Whilst the animation was my induction, there were two previous incarnations, the printed graphic novel (Briggs, 1982), and on BBC Radio 4 *The Monday Play* (1983). In an interview, Briggs and the radio producer shared intriguing insights. The producer expressed (now proven incorrect) doubts about rendering WTWB in the style of *The Snowman*. However, what fascinated me was Briggs' belief that the story would not influence the government. Conversely, it was discussed twice in Parliament: the novel received praise in the House of Commons. Additionally, Lord Jenkins of Putney sought assurance from Baroness Hooper that it wouldn't be banned in schools (Rogers, 2018). Briggs received correspondence from individuals worldwide describing the profound influence of WTWB on their lives. I continue to contemplate this work due to its profound nature. It demonstrates that the most impactful works of art often challenge our assumptions and compel us to confront uncomfortable truths.

A segment of an interview with director Jimmy T. Murakami provides testimony regarding his personal motivations which provides clarity ('When The Wind Blows – Interview with Director Jimmy T. Murakami', 2020). He is Japanese, and his family members suffered during the end of WWII. He directed the animation to ensure young people (particularly his daughters) understood what is at risk. I realised that whilst WTWB is a fictional narrative of quintessential English construction, the consequences are not an abstract 'what if' but were experienced by the citizens of Hiroshima and Nagasaki, creating intergenerational trauma. The production sat in my mind, previously viewed through my naive childhood lens, but this new context sent shivers down my spine. Briggs expressed disappointment with the cheerful ending of the 'Snowman', indicating that his original intention was to teach children about the concept of death (Giles, 2016). Yet with WTWB, I think he succeeded in creating a thought-provoking, poignant and terrifying narrative that left a lasting impression on many. Curious to discover if this production still stalks other people's subconscious: 'When The Wind Blows – Nuclear Snowman' (2017) is a dark comedic review that mirrors my perceptions: 'If you fancy crying this Christmas, and you've only got forty-five minutes to make that happen, this is your movie.'

17.4.4 Popular Fiction

Briggs sparked my interest, concern, unease, and paranoia regarding nuclear technology management. I value the distinctly British perspective embedded in his work. Nonetheless, I am confident that international readers can relate to alternative cultural touchpoints, such as the post-apocalyptic novel *Z for Zachariah* by O'Brien (1987) and *The Simpsons* (1990). In the iconic series, Homer, at the nuclear power plant, comically fumbles with radioactive rods during the opening sequence, complete with atomic goo dripping from the ceiling. Initially, I amusingly thought nuclear reactor operations were solely focused on safety, not realising the significant impact of the Three Mile Island incident on American society and policy at the time. This iconic introduction and the narrative of Homer's employment at the nuclear plant were ever-present during my childhood and further cemented my intrigue in radiological matters. Unsurprisingly, video essays dissecting Homer's job exist: 'Homer Simpson is the Worst Nuclear Technician Ever' (2022).

17.5 Radio-Ethnography

Five significant disasters involving nuclear power have contributed to my fascination. These individual events are intensely personal to each nation, which aims to mitigate the perception of danger just as much as the actual hazard. To help contextualise these events, we can consider four metrics: A) the amount of radiation leaked into the environment, B) the number of resultant fatalities, C) the economic cost, and D) the INES rating, which ranges from 1 (anomaly) to 7 (major incident), as defined by the International Atomic Energy Agency (2008). Sovacool et al. (2016), and Wheatley et al. (2017), discuss the incidents in the context of economics and the environment. Smythe (2011) presents arguments regarding the validity and accuracy of the INES scale used to quantify these events (Table 17.1). These metrics should be considered approximate at best; for example, the measurement 'TBq': named after Becquerel (The Nobel Prize in Physics, 1903), indicates the amount of radiation released but does not specify the radiation type. Different isotopes are more harmful and longer lasting than others. Windscale/ Sellafield is notably prevalent in the data, a trend Smythe (2011) attributed to meticulous record

TABLE 17.1 The INES ratings

Location	Year	TBqs	INES	Economic Cost USD	Direct Fatalities
Windscale, UK	1957–1984	3,925~	5	17,129 million	~0
Kyshtym, Soviet Union	1957	1,000,000	6	1733.4 million	103
Three Mile Island, USA	1979	3,700,000	5	2773.4 million	~0
Chernobyl, Ukraine	1986	5,200,000	7	7742.5 million	4056
Fukushima, Japan	2011	630,000	7	162,650.7 million	573

keeping, a higher frequency of accidents, and a general underreporting of minor events in the global industry.

17.5.1   1957: Explosive Secrets

A significant fire broke out in the graphite moderator of the Windscale (UK) reactor, marking the first widely publicised release of radioactive substances into the environment (Bergan et al., 2008). The fire consumed eight tonnes of uranium and required two million gallons of water to cool the core. It was revealed years later that radioactivity had been leaking for a considerable time before the disaster. Due to the unsanctioned persistence of Dr. Leslie and Dr. Jakeman, who commandeered a GMC and discovered radioactive particles in their gardens, prompting action from the government, the situation may have unfolded differently (Our Reactor Is on Fire, 1990). Contrary to what many were told then, Windscale's primary function wasn't electricity generation; it consumed more power than it produced (Russell, 2022). Its actual role was to enrich uranium for the atomic programme, demonstrating to the US that the UK had comparable nuclear capabilities.

Post-incident, the IAEA advised, 'Don't drink the milk, but everything else is fine' (Loutit et al., 1960). I viewed with suspicion the 1981 rebranding of the 'Windscale' site to 'Sellafield', a move surely designed to obscure its controversial history (Corecumbria, 2023). An incident occurred recently in 2005, rated 3 on INES (Nuclear Accidents Worldwide, 2024). While the UK was managing the events at Windscale, a significant incident occurred at Kyshtym in the Soviet Union. It remained secret for decades but was equally impactful. The human suffering during construction and the radiological contamination from the incident were concealed at the time and are still being measured. Akleyev et al. (2017) highlight the differences in data interpretation and collection; interestingly, the Soviets were concerned not only about the radiation levels in the milk but also the wheat: a notable difference in the IAEA guidance during Windscale.

17.5.2   The '80s

As a child, I misunderstood the Three Mile Island disaster as occurring on an island three miles offshore. Later, realising it had happened near a populated urban area, I felt a mix of amusement over my naivety and horror. The incident still inspires many texts and documentaries. *Meltdown: Three Mile Island* (2022) offers an emotive narrative from the personal perspectives of Rick Parks and others involved. The 1986 BBC *Newsnight* (2019) disaster coverage and the concern on my mother's face made me realise the significance. Increased radioactivity was recorded in the Nordic countries before a formal announcement was made (Higley, 2006). The disaster and the culture that precipitated it and hindered response efforts continue to fascinate many: The *Chernobyl Mini Series* (2019) offers a sobering view of the implications of systemic mismanagement despite debates on its technical accuracy. A section of the narrative reveals that the GMCs being used were all reporting the same levels, attributed to the now infamous line 'not great, not terrible'. The horrendous truth was that due to the inherent design limitations, the GMCs they had access to could not interpret levels above a maximum frequency (Nyquist, 1928).

17.5.3   Recent Events

The control room engineers in the Fukushima Daiichi power station experienced 'over-sonification'; the plethora of alarms simultaneously sounding made prioritising the response challenging. The fundamental difference between Fukushima and Chernobyl is that the former was caused by nature, and the latter was man-made (Perko et al., 2019). The repercussions of this incident are still felt, not just in Japan, where clean-up will continue for decades, but globally, influencing policy decisions worldwide. For instance, Germany increased its reliance on coal after decommissioning its nuclear power capacity, a change initiated with environmental concerns in mind following Fukushima but with unclear long-term benefits (The nuclear phase-out in Germany, 2023).

17.6 Personal Experience

17.6.1 Bunkers

A key moment that elevated my conception of the wider world was visiting family in Switzerland in the early 1990s. I discovered that the Swiss take nuclear planning seriously. Each apartment was assigned a storage area in the basement of my father's residence. The most robust residential door I have ever seen was at the storage entrance. It was colossal, square, and spectacularly thick. Mounted on enormous hinges, giant nuts and bolts would seal the occupants inside during fallout. My father insisted it would never be used, but I could not help contemplating its construction motivations. Are the Swiss paranoid, or are the UK complacent? Since 1963, residential construction rules in Switzerland have required a shelter (Mariani, 2009). It was amended in 2012 for smaller developments, allowing participation in the community shelter provision (Copley, 2011). Recent events in Ukraine have shifted the dialogue to emphasise national readiness and coverage (Lam, 2022, Carney, 2022).

17.6.2 The Ukrainians

Shortly after the start of the Russian invasion of Ukraine (Putin launches full-scale invasion, 2022), we decided as a family to open our home to a Ukrainian mother and her infant son. The

FIGURE 17.2   A smaller residential bunker door. The spanner is bright yellow.

decision was simple: a family from Kyiv needed shelter, we had space, so we helped. This choice fundamentally changed us in ways difficult to articulate, and its impact on their family is beyond my speculation. From the moment I collected them at the airport arrivals, a profound chain of events was now in motion. I witnessed our incoming mother, arms wrapped protectively around her son as if trying to absorb him, shielding him from the world, with a look that told a tale of a long, arduous, complex journey. She was fleeing a warzone, leaving her husband, parents, pets, and everything she called home, travelling selflessly into an uncertain future, an unfamiliar land, to be hosted by a strange family.

Having actual Ukrainians living with us compelled me to educate myself. As our guests shared updates about their family's movements within Ukraine, my interest shifted from being a concerned citizen to something deeply personal. Most relationships can be easily defined and categorised with words like mother, brother, daughter, friend, lover, neighbour, colleague, compatriot, or enemy. However, our connection went beyond any conventional definition. It was a complex blend of duty, care, and a commitment to preserving their privacy and independence.

Reflecting on Russia's 2014 Crimea annexation, I was initially reassured by reports of the population being primarily Russian speakers. I didn't realise then that this action violated agreements, including those the UK signed. The titles alone of the United Nations Treaty Series (2014) highlight undeniable failures in the foreign policies of both the UK and the USA.

1. 'No. 52240. Ukraine and Russian Federation: Treaty on Friendship, Cooperation, and Partnership between Ukraine and the Russian Federation. Kiev, 31 May 1997'.

   At this juncture, the notion of the Russian Federation extending friendship towards Ukraine seems bitterly ironic.
2. 'No. 52241. Ukraine, Russian Federation, United Kingdom, and United States of America: Memorandum on security assurances in connection with Ukraine's accession to the Treaty on the Non-Proliferation of Nuclear Weapons. Budapest, 5 December 1994'.

Known as the Budapest Memorandum, it extended security guarantees to a post-Soviet Ukraine, encouraging the country to decommission the world's third-largest nuclear arsenal (Budjeryn, 2022). Reflecting on the events of 2014, it's impossible not to wonder whether upholding this agreement could have led to a more stable geopolitical landscape. It may have averted yet another broken promise weighing heavily on our collective conscience. It is vital to highlight that Chernobyl is in Ukraine, not Russia, a seemingly obvious fact somewhat blurred before the dissolution of the

FIGURE 17.3 The new arrivals.

Soviet Union. I recall struggling to acquaint myself with the reshaped geopolitical and cartographic realities.

17.6.3 Can 'I' Measure It?

Currently, two instances in the Ukraine war have raised concerns about radiological incidents. The first occurred when Russian troops dug trenches in the 'red forest' at Chernobyl, seemingly unaware of the soil contamination (Reuters, 2022). The second occurred during the occupation of Zaporizhzhia and its nuclear plant. The news reports were contradictory regarding the control and the state of the situation. Investigating the fundamental differences between the Chernobyl and Zaporizhzhia reactor designs was reassuring, as the latter shares none of the flaws of the former (Winfrey, 2022). During a conversation with a friend about the situation, we shared concerns that the Russian state might somehow exploit a nuclear disaster. However, neither could fully contemplate the likelihood or consequences. We deliberated when, during an intended or unintended incident, how soon the public would learn about it, considering previously discussed scenarios.

17.6.4 Ethno-Conclusion

A unique convergence of cultural narratives and personal experiences sparked my foray into measuring radiation and its creative application. The stark portrayal of nuclear realities in Raymond Briggs's *When the Wind Blows* first captured my imagination. This, combined with early misconceptions about nuclear politics during the Reagan-Gorbachev era and later insights from hosting Ukrainian refugees, directed my attention to the tangible aspects of nuclear technology. Exposure to Switzerland's proactive nuclear planning further heightened my interest. These diverse influences collectively ignited a fascination with the practicalities of radiation measurement, leading to this journey that marries scientific concepts with artistic creation. Inspired by Dr Frank Leslie's pragmatic approach (he took an unauthorised GMC to his home and dug up a particle of uranium oxide; he still has it, and it is still highly radioactive) along with the tapestry of influences and events detailed earlier, the presented narrative is a manifestation of my journey navigating the conflicting duplicitous nature of nuclear power and weapons and considering my own biases: I trust the technology, just not the humans managing it. The presented synthesis of fiction and reality reveals a complex relationship between the populous and the nuclear industry. Radiological technology is deployed in too many areas of modern civilisation to discuss; the use sits across the spectrum from healing to destruction. We should also be circumspect that even in the best-case scenario, we should be prepared for a 'normal accident', a product of complex systems and human error (Perrow, 1999).

17.7 Exegesis

17.7.1 Sonification vs the Apocalypse

Scaletti (2018) coined and defined the term sonification as an activity with an inherent scientific purpose: to examine data and reveal fresh perspectives. Andreopoulou and Goudarzi (2021) demonstrate that whilst an artistic component is becoming more common, the scientific purpose should be demonstrated for the activity to be considered sonification. The presented artworks are collectively titled 'The Sonification of the Apocalypse'. Whilst an aspirational goal of MaxVCS is to provide a conduit for sonification for the direct purposes of science, this work does not

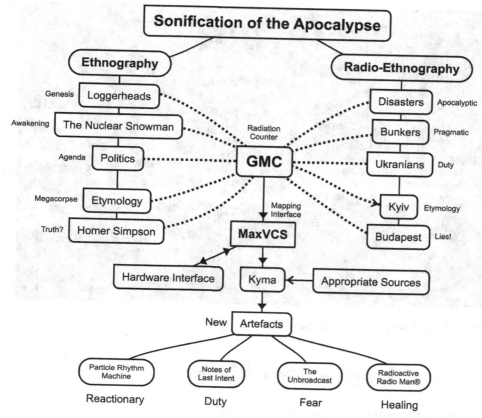

FIGURE 17.4 Mind map of fictional and actual events of influence.

fully embody the definition of sonification. Or does it? If we allow ourselves a little latitude from Scaletti's strict definition, the GMC itself is a 'sonification device'; therefore, the ping it produces, and anything triggered by it, is a sonification of the radiation at that location and time. The apocalyptic element of the title intentionally stretches and plays with the definition, yet how will we know if the apocalypse is upon us? Perhaps it has already started? Not according to *my* measurements.

17.7.2 Geiger Counter In The Studio

The GMC outputs a short, low-voltage pulse. The circuit was modified to output the pulse from the cathode tube before the A-D and microcontroller.

A series of Eurorack modules is used to convert, buffer, lengthen, and step down the voltage so that the Arduino would read the 'pings' reliably.

17.7.3 The Three Tools

Despite offering a single pulse at a random time interval, three techniques can be employed to harness the motion for (perhaps) musical purposes:

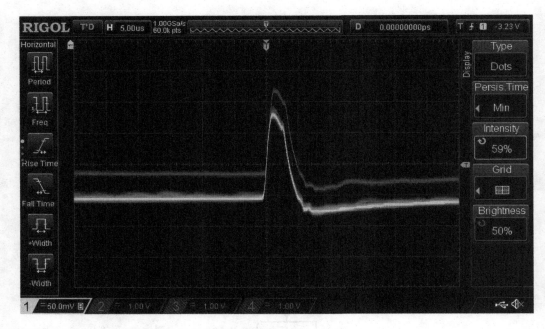

FIGURE 17.5 An oscilloscope plot of a single GMC ping.

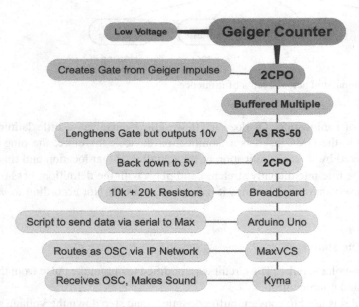

FIGURE 17.6 Topology of GMC to Kyma vs MaxVCS.

17.7.3.1   The Basic Ping

As we have established, the GMC produces a series of triggers at unpredictable times, and sometimes in clusters. Once digitised, the output is binary. An example: '110100100'. The first obvious and perhaps mundane step in this exploration is triggering a sound each time the GMC registers a particle. This randomly distributed event is distinctly un-musical as it has no inherent rhythm, velocity, or pitch. It can be harnessed to generate curious, sparse ambient textures.

17.7.3.2   Count the Time

The second iteration is to measure the time between the pings. This can be divided or multiplied in both analogue and digital domains. This next step unlocks a range of potential applications. The value produced can be used as a BPM, delay time, random seed, and anywhere else a value can be mapped. The output is a series of integers, for example, '1, 836, 5, 7, 5830, 14, 35, 70, 2, 4030, 10, 7192'.

17.7.3.3   Smooth the Result

The third iteration introduces a second variable, a time value applied as a smoothing function (linear or non-linear), creating a stream of peaks and troughs. This can be used to change parameters smoothly. This is analogous to the 'Time to Voltage Converter' Ahmed (2007) discusses.

This output produces unexpectedly unique motions that can inject dynamism into a composition, although there is a caveat: to ensure the output ranges correctly, it does require a threshold to be set, and consequently, a minimum number of pings need to arrive in a given timeframe, or the control flatlines.

These three techniques for extracting different data patterns from the radiation can be used individually or in combination. Considering the definitions presented by Dean and McLean (2018): the GMC techniques can be thought of in the context of the 'non-human' musician; whilst the radiological source itself does not exhibit any 'intelligence', algorithms can be built to be 'reactive' and contribute to creating chaotic motion that is constrained enough for the brain to impose patterns upon. Effects balances, detuning, delay times, shaping amounts, panning modulation, and many other combinations have yielded fascinating results.

17.7.4   The Machine That Goes Ping

The 'Proof of Geiger OSC Kyma and Volta Concept' (2022) demonstrates the first and most straightforward implementation, additionally serving as a proof-of-concept for MaxVCS. The background-radiation ping is routed to Kyma in addition to Volta Create (2023) software from MaxVCS: enabling cross-domain real-time visual generation. Scaletti (2022) commented on the

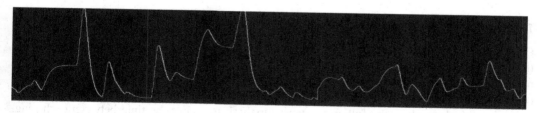

FIGURE 17.7   A time-to-voltage plot.

FIGURE 17.8 Proof of ping.

FIGURE 17.9 Improvisation for polyphonic synthesiser and Geiger Müller radiation counter in Kyma.

piece: "This is one time when we welcome sparse textures!" All can relate to her reaction, as the implications of an increased tempo of readings are difficult to comprehend. This piece is a subtle infusion of traditional performance modulated by the passing radioisotopes – a meditation upon the status quo.

17.7.5 Radioactive Effects Processing

'Improvisation for Polyphonic Synthesiser and Geiger Müller Radiation Counter in Kyma' (2023) also demonstrates modulation generated by a peaceful background radiation. The display reassuringly does not register anything but 'safe'. The synthesiser is a Cold War era analogue Oberheim Matrix 1000 processed through seven different blended effects chains. Comb filtering, bit-crushing, granulation, phasing, delays, filtering, and reverbs are all selected and balanced by the GMC pings and the algorithmic responses.

17.7.6 Particle Rhythm Machine

Cook (2017) sets out thirteen principles for interface design, two are relevant when considering this piece; 'that new algorithms suggest new controllers', and conversely, 'new controllers suggest new algorithms'. This piece developed in sympathy with constructing a simple hardware interface (Figure 17.9). The GMC pings are interpreted using a combination of the previously mentioned strategies, which result in the following steps: Firstly, it measures the time between each ping, then calculates a BPM value, but crucially waits until the next beat (defined by the previous ping) to change to the new value. This constantly shifting but rhythmically quantised BPM drives a percussion and melodic sequencer and instantly changes the pitch of looped material to match the BPM. Procedural limits were placed on the upper and lower BPM values: Too low, it doubles. Too fast, it halves (Particle Rhythm Machine Techno Experiment Kyma, 2023). The hardware interface completely changed the creative relationship with the sound structure. Performing with a physical toggle switch has a satisfying, robust, and industrial feeling, as though they are connected to vital systems in a space shuttle or even a nuclear reactor control panel. The haptic feel is discussed in

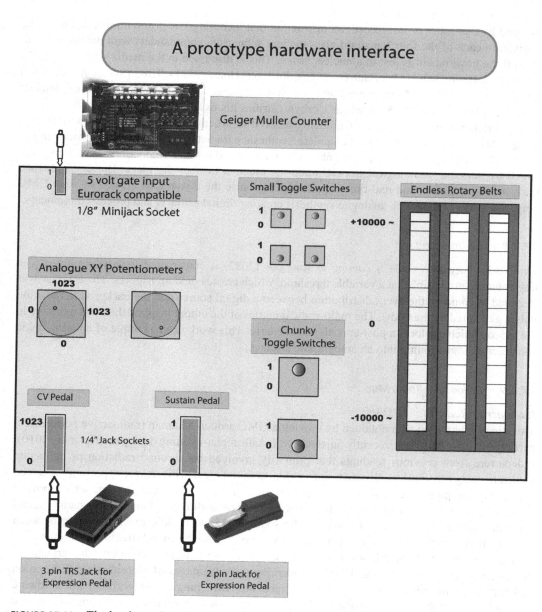

FIGURE 17.10 The hardware interface layout.

many texts, but the humble toggle has much to offer: The state can be observed through touch alone, it is haptic by nature, and the control state is, in effect, 'memory', as the position remains permanent regardless of power interruption. The XY analogue controllers feel snappy and precise due to the increased resolution of MaxVCS from 7 bits to 11 bits (0–127 vs 0–1023). A wide range of integrations of the integers and the smoothed streams are harnessed to modulate various parameters within the sound structure.

17.7.7 The Notes of Last Intent

The Notes of Last Intent (2023)
Inspired by the 'letter of last resort', an instruction that incoming UK prime ministers must write is given to each of the four Vanguard-class nuclear submarine commanders with details on what to do if the government is lost to a nuclear strike. This is discussed in the media when a new PM is installed: Blair (2006), May (Garrett, 2016), Johnson (Hoare, 2021), Truss (Sabbagh, 2022) and Sunak (Reporter, 2022). Most interesting is the Scottish paper 'Corbyn: I would never use nuclear weapons' (The Herald, 2015) in which Corbyn clarifies his exact intentions. This, in the eyes of some, derails the whole concept of mutually assured destruction. The composition was created using subtle GMC modulation to manipulate synthesis parameters before the final processing in Kyma. The Geiger pings were elevated above the background by the introduction of uranium glass, which triggers the 'radiological impact' upon the creative flow of the piece. Bursts of glitched pitch-modified buffers and real-time time-stretching cause the disruptions created by each GMC ping. The pings collide with analogue synthesis creating digital artifacts and processed moments.

17.7.8 The Un-Broadcast

The Un-Broadcast: No one is coming to save you (2023) is constructed from white noise. The noise sources are sampled at a variable threshold, which creates random triggers. The exercise was intended to compare the event distribution between a digital source and the background radiation pulses generated by the GMC. The radio-static qualities of the output inspired the short narrative in the video, which explores a post-apocalyptic scenario. This work is an example of an allographic construction transitioning into an autographic artefact.

17.7.9 Radioactive Radio Man

Radioactive Radio Man, 2023
This piece is based on data obtained by moving a GMC around a human (radioactive performer?) participant who had been recently injected with safe alpha-emitting isotopes (Holzner, 2010), a departure from previous readings that primarily involved background radiation or deliberate exposure using nominally radioactive sources like uranium glass. This unique opportunity allowed for examining particle emission levels, density, and distance from the subject, which proved captivating and instrumental in creating this piece. The audio functions as a dynamic sound sculpture, rapidly transitioning between audio sources, akin to a radio tuner switching between stations. It utilises the frequency of particle counts to modulate the sound structure continuously.

These synthesised techniques and concepts draw direct inspiration from the arguments presented by Dean and McLean (2018) concerning the advantages of algorithmic collaboration and authorship in artistic expression. In this piece, the GMC pings, the algorithm, and hardware interface manifest this piece. The distribution of the triggers was fascinatingly different to the sparse

FIGURE 17.11   Particle rhythm machine demonstration.

FIGURE 17.12   Notes of last intent.

FIGURE 17.13   The Un-Broadcast (YouTube).

background readings and even compared to the count produced by the uranium glass. The GMC display changed from 'safe' to 'unsafe', and then to 'danger' rapidly when in proximity. Alpha particles are not harmful and do not travel far or through dense matter. I considered the horror of measuring harmful radiation after an incident or detonation. This GMC cannot distinguish between the different isotopes, and I contemplated purchasing a more sophisticated model. However, this piece provides a rational and paranoia-dissolving conclusion, exploring the benefits of radiation in motion and interaction with a healing process. It is important to note that this piece warrants an ethical declaration. Ethical considerations were regrettably overlooked in the excitement of exploring the GMC's potential with an alternative radiation source. The human subject, who

FIGURE 17.14 Radioactive Radio Man (YouTube).

FIGURE 17.15 nortotron.co.uk

approached the measurements with curiosity and grace, deserves heartfelt gratitude for enabling the collection of such intriguing data. It has become clear that medical investigations of this nature are not undertaken without significant purpose, a lesson in humility.

17.8 Conclusion

Experimenting with the GMC pulses and the unique hardware interface has demonstrated that any voltage can be taken and routed using MaxVCS to Kyma and harnessed for creative intent.

Paranoia is perhaps a more unusual creative force, although curiosity is an equal factor in this endeavour. The combined narrative elements have all etched an indelible mark upon my consciousness, fostering a nuanced synthesis of truth that bridges the gap between the fictional and the historical narratives. It serves as a poignant reminder that while fictional sources sometimes emanate genuine sincerity and truth, the information disseminated by governments and corporate entities is often modulated, usually revealing just the tip of the iceberg and only when it becomes conspicuously undeniable. The artworks have manifested through curiosity, rule-breaking, paranoia, and nostalgia whilst also serving as an artistic proof of concept for the development of MaxVCS. This chapter has traversed personal landscapes that resonate with broader cultural rhythms and narratives. Touching upon distinct cultures' attitudes and the nuanced impact of fictional works, drawing upon the profound creations of Raymond Briggs. We have further delved into an analysis of significant radiological occurrences and the complex narratives surrounding them, often orchestrated by governing bodies and corporations with interests that occasionally supersede safety concerns. Described as 'Normal Accidents' by Perrow (1999), it is the concept of inevitable accidents occurring in complex technological systems. This is where the 'subtle activism' of this research emerges, nuclear disarmament is far too complex to comprehend and deconstruct, but there is merit in Mr Parks' statement in *Meltdown: Three Mile Island* (2022); he suggests that the nuclear power industry should be run and managed by the military on a national scale with corporate interests removed. We have embarked on a reflective journey through personal and cultural resonances that fuelled this endeavour. As a result, a unique combination of technologies and creative mapping strategies have emerged, facilitating a deep dive into the unseen and intangible radiation around us, which nonetheless leave measurable imprints ripe for exploration and interpretation in our quest for authentic engagement, understanding, and empathy with the world surrounding us.

To learn more about MaxVCS please visit nortotron.co.uk

References

BBC Newsnight (2019) 1986's Chernobyl disaster.. Available at: www.youtube.com/watch?v=ET6ov0Kv p1M (Accessed: 1 August 2023).

A Grand Day Out (1994). Aardman Animations, National Film and Television School (NFTS), Polyphony Digital.

Adams, T.E., Holman Jones, S.L. and Ellis, C. (eds) (2022) *Handbook of autoethnography*. Second edition. New York, NY: Routledge, Taylor & Francis Group.

Ahmed, S.N. (2007) *Physics and engineering of radiation detection*. First edition. Amsterdam; Boston: Academic Press.

Akleyev, A.V. *et al.* (2017) 'Consequences of the radiation accident at the Mayak production association in 1957 (the 'Kyshtym Accident')', *Journal of Radiological Protection*, 37(3), p. R19. Available at: https://doi.org/10.1088/1361-6498/aa7f8d.

An Ode to Prototyping: The Adventures of an Audio Nerd: Breadboard, Shield, Strip-Board, & Arduino (2023). Available at: www.youtube.com/watch?v=8NE3M8PCpxo (Accessed: 8 September 2023).

Andreopoulou, A. and Goudarzi, V. (2021) 'Sonification First: The Role of ICAD in the Advancement of Sonification-Related Research', in *Proceedings of the 26th International Conference on Auditory Display (ICAD 2021)*. *ICAD 2021: The 26th International Conference on Auditory Display*, Virtual Conference: International Community for Auditory Display, pp. 65–73. Available at: https://doi.org/10.21785/icad2021.031 (Accessed: 2 August 2023).

BBC TV News (2020, Jan) Coronavirus: first BBC TV News mentions. Available at: www.youtube.com/watch?v=jE7yTbPKZq8 (Accessed: 3 August 2023).

Bergan, T., Dowdall, M., & Selnæs, Ø. G. (2008). On the occurrence of radioactive fallout over Norway as a result of the Windscale accident, October 1957. *Journal of Environmental Radioactivity*, 99(1), pp. 50–61.

Blair, T. (2006) *Exchange-of-letters-between-the-Prime-Minister-and-the-President-of-the-United-States-of-America.pdf*. Available at: www.nuclearinfo.org/wp-content/uploads/2020/09/Exchange-of-letters-between-the-Prime-Minister-and-the-President-of-the-United-States-of-America.pdf (Accessed: 13 May 2023).

Briggs, R. (1982) *When the wind blows*. UK: Penguin Books.

Budjeryn, M. (2022) 'The breach: Ukraine's territorial integrity and the Budapest memorandum'. Available at: www.wilsoncenter.org/publication/issue-brief-3-the-breach-ukraines-territorial-integrity-and-the-budapest-memorandum (Accessed: 4 August 2023).

Carney (2022) *Ukraine conflict prompts Swiss to check their bunkers*. Available at: https://newseu.cgtn.com/news/2022-05-08/Switzerland-only-country-with-shelter-space-for-everyone-19RLO52Pbhe/index.html (Accessed: 4 August 2023).

Chernobyl Mini Series (2019). Available at: www.imdb.com/title/tt7366338/?ref_=nv_sr_srsg_0_tt_8_nm_0_q_chernob (Accessed: 4 September 2023).

Cook, P. (2017) '2001: Principles for designing computer music controllers', in A.R. Jensenius and M.J. Lyons (eds) *A NIME Reader*. Cham: Springer International Publishing (Current Research in Systematic

Musicology), pp. 1–13. Available at: https://doi.org/10.1007/978-3-319-47214-0_1 (Accessed: 14 August 2023).

Copley, C. (2011) 'Swiss relax nuclear shelter construction law', *Reuters*, 1 December. Available at: www.reuters.com/article/us-swiss-bunker-idUSTRE7B01RP20111201 (Accessed: 24 August 2023).

The Herald. (2015) Corbyn: I would never use nuclear weapons. Available at: www.heraldscotland.com/news/13793287.corbyn-never-use-nuclear-weapons/ (Accessed: 18 September 2023).

Corecumbria (2023) Available at: http://corecumbria.co.uk/alternative-tour-of-sellafield/sellafield/ (Accessed: 14 April 2023).

Cycling 74 (2020) *What is Max? | Cycling '74*. Available at: https://cycling74.com/products/max (Accessed: 12 April 2023).

Dean, R.T. and McLean, A. (eds) (2018) *The Oxford handbook of algorithmic music*. New York, NY: Oxford University Press (Oxford handbooks).

etymonline.com (2023) Megadeath (n.). Available at: www.etymonline.com/search?q=megadeath (Accessed: 21 August 2023).

Garrett, M.G. (2016) 'Theresa May's grim first task: Preparing for nuclear armageddon', *POLITICO*, 15 July. Available at: www.politico.eu/article/the-grim-task-awaiting-theresa-may-preparing-for-nuclear-armageddon-uk-prime-minister-british-defense-letter-of-last-resort/ (Accessed: 3 August 2023).

Giles, L. (2016) The shocking truth behind Raymond Briggs' The Snowman | Decor Books. Available at: https://decorbooks.co.uk/uncategorized/the-shocking-truth-behind-raymond-briggs-the-snowman/ (Accessed: 20 August 2023).

Gonot-Schoupinsky, F. (2022) 'Pragmatic autoethnography – A brief idea in Journal of Brief Ideas August 31 2022'. Available at: https://doi.org/10.5281/zenodo.7064069, www.researchgate.net/publication/363405577_Pragmatic_Autoethnography_-_A_Brief_Idea_in_Journal_of_Brief_Ideas_August_31_2022 (Accessed: 11 August 2023).

Google Map. Loggerheads is a village which sits at the intersection of Staffordshire, Shropshire, and Cheshire in the United Kingdom. www.google.com/maps/place/Loggerheads,+Market+Drayton,+UK/@52.9208502,-2.3962785,15z/data=!3m1!4b1!4m6!3m5!1s0x487a6230a4dafb41:0xb9b6e72f52e0fd4b!8m2!3d52.919143!4d-2.389342!16zL20vMDU4OV96?entry=tt

Gullo, T. and Hepworth-Sawyer (2024) *Innovation in music: Technology and creativity*. London: Routledge.

Higley, K.A. (2006) 'Environmental consequences of the Chernobyl accident and their remediation: twenty years of experience. Report of the Chernobyl forum expert group "environment"', *Radiation Protection Dosimetry*, 121(4), pp. 476–477. Available at: https://doi.org/10.1093/rpd/ncl163 (Accessed: 11 July 2023).

Hoare, C. (2021) Boris Johnson's 'letter of last resort' exposed amid doomsday threat, Express.co.uk. Available at: www.express.co.uk/news/science/1389757/boris-johnson-nuclear-warning-letter-doomsday-clock-stream-world-war-3-coronavirus-spt (Accessed: 10 July 2023).

Holzner, S. (2010) *Physics II for dummies*. Hoboken, NJ: Wiley (--For dummies).

Homer Simpson is the Worst Nuclear Technician Ever (2022). Boris Johnson's 'letter of last resort' exposed as Doomsday Clock risks midnight strike. Available at: www.youtube.com/watch?v=3D--jytkUW8 (Accessed: 25 April 2023).

@frequencymanipulator (2023) Improvisation for polyphonic synthesiser and Geiger muller radiation counter in Kyma. Available at: www.youtube.com/watch?v=Ruv-P6ileow (Accessed: 8 March 2023).

International Atomic Energy Agency (2008) 'The International Nuclear and Radiological Event Scale'. www.iaea.org/resources/databases/international-nuclear-and-radiological-event-scale

Knoll, G.F. (2000) *Radiation detection and measurement*. Third edition. New York: Wiley.

Lam, C. (2022) Swiss bomb shelters are readied as international tensions mount, euronews. Available at: www.euronews.com/2022/04/03/nuclear-bunkers-for-all-switzerland-is-ready-as-international-tensions-mount (Accessed: 11 July 2023).

Lee, G. (1992) 'Arriving Gorbachevs welcomed by Reagans', *Washington Post*, 3 May. Available at: www.washingtonpost.com/archive/politics/1992/05/03/arriving-gorbachevs-welcomed-by-reagans/c441b525-793a-4afb-b613-d0c843b068a4/ (Accessed: 3 July 2023).

Loutit, J.F., Marley, W.G., & Russell, R.S. (1960) The nuclear reactor accident at Windscale - October, 1957: Environmental aspects (INIS-XA-N--258). International Atomic Energy Agency (IAEA). http://inis.iaea.org/search/search.aspx?orig_q=RN:37004435

Mariani, D. (2009) Bunkers for all, SWI swissinfo.ch. Available at: www.swissinfo.ch/eng/prepared-for-anything_bunkers-for-all/995134 (Accessed: 22 July 2023).

Meltdown: Three Mile Island (2022). Moxie Pictures, Netflix Studios.

merriam-webster.com (2023) *Definition of MEGADEATH*. Available at: www.merriam-webster.com/dictionary/megadeath (Accessed: 13 July 2023).

ncsc.gov.uk (2018) Available at: www.ncsc.gov.uk/news/russian-military-almost-certainly-responsible-destructive-2017-cyber-attack (Accessed: 22 July 2023).

Nuclear Accidents Worldwide (2024) *Statista*. Available at: www.statista.com/statistics/273002/the-biggest-nuclear-accidents-worldwide-rated-by-ines-scale/ (Accessed: 9 January 2024).

Nyquist, H. (1928) 'Certain topics in telegraph transmission theory', *Transactions of the American Institute of Electrical Engineers*, 47(2), pp. 617–644. Available at: https://doi.org/10.1109/T-AIEE.1928.5055024 (Accessed: 12 July 2023).

O'Brien, R.C. (1987) *Z for Zachariah*. NYC, USA: Simon and Schuster.

Our Reactor Is on Fire (1990) *Inside story*. Available at: www.youtube.com/watch?v=vcsyMvQtlKs (Accessed: 4 July 2023).

Particle Rhythm Machine Techno Experiment Kyma (2023). Available at: www.youtube.com/watch?v=GTiYRgCNVO0 (Accessed: 11 September 2023).

Perko, T. *et al.* (2019) 'Fukushima through the prism of Chernobyl: How newspapers in Europe and Russia used past nuclear accidents', *Environmental Communication*, 13(4), pp. 527–545. Available at: https://doi.org/10.1080/17524032.2018.1444661 (Accessed: 04 July 2023).

Perrow, C. (1999) *Normal accidents – Living with high risk technologies – updated edition*. Updated edition with a new afterword and a new postscript by the author. Princeton, NJ: Princeton University Press.

Proof of Geiger OSC Kyma and Volta Concept (2022). Available at: www.youtube.com/watch?v=oajvk-Nje90 (Accessed: 11 September 2023).

Radioactive Radio Man (2023). Available at: www.youtube.com/watch?v=WFglg8MHzpM (Accessed: 11 September 2023).

Reporter, L. political (2022) Sunak will have an intelligence briefing on 'imminent threats' to UK and will 'write letters of last resort' amid Putin's nuclear threats, London Business News | Londonlovesbusiness.com. Available at: https://londonlovesbusiness.com/sunak-will-have-an-intelligence-briefing-on-imminent-threats-to-uk-and-will-write-letters-of-last-resort-amid-putins-nuclear-threats/ (Accessed: 1 July 2023).

Reuters (2022) Unprotected Russian soldiers disturbed radioactive dust in Chernobyl's "Red Forest", workers say, 29 March. Available at: www.reuters.com/world/europe/unprotected-russian-soldiers-disturbed-radioactive-dust-chernobyls-red-forest-2022-03-28/ (Accessed: 8 January 2023).

Roads, C. (1996) *The computer music tutorial*. Cambridge, Massachusetts, USA: MIT Press.

Rogers, J. (2018) 'Einstein's monsters', *New Statesman*, 147(5410), pp. 48–52.

Russell, P. (2022) Windscale, Keele University. Available at: www.keele.ac.uk/extinction/controversy/windscale/ (Accessed: 14 July 2023).

Russia Ukraine conflict: Putin launches full-scale invasion (2022). Available at: www.youtube.com/watch?v=aCOekp64He8 (Accessed: 8 April 2023).

Sabbagh, D. (2022) 'Letters of last resort: PM's early task to write to UK's nuclear sub commanders', *The Guardian*, 6 September. Available at: www.theguardian.com/uk-news/2022/sep/06/letters-of-last-resort-pms-early-task-to-write-to-uks-nuclear-sub-commanders (Accessed: 13 May 2023).

Scaletti, C. (2018) 'Sonification ≠ music', in R.T. Dean and A. McLean (eds), The Oxford Handbook of Algorithmic Music. Oxford University Press. Available at: https://doi.org/10.1093/oxfordhb/9780190226992.013.9, www.researchgate.net/publication/326414925_Sonification_music (Accessed: 15 April 2023).

Scaletti, C. (2022) 'Carla communications'. Discord, Norton, C, 8th Dec.

Shampo, M.A., Kyle, R.A. and Steensma, D.P. (2011) 'Hans Geiger—German physicist and the Geiger counter', *Mayo Clinic Proceedings*, 86(12), p. e54. Available at: https://doi.org/10.4065/mcp.2011.0638 (Accessed: 15 April 2023).

Smythe, D. (2011) 'An objective nuclear accident magnitude scale for quantification of severe and catastrophic events', *Physics Today* [Preprint]. Available at: https://doi.org/10.1063/PT.4.0509 (Accessed: 10 August 2023).

Sovacool, B.K. *et al.* (2016) 'Balancing safety with sustainability: assessing the risk of accidents for modern low-carbon energy systems', *Journal of Cleaner Production*, 112, pp. 3952–3965. Available at: https://doi.org/10.1016/j.jclepro.2015.07.059 (Accessed: 22 April 2023).

Symbolic Sound (2023). Available at: https://kyma.symbolicsound.com/ (Accessed: 11 September 2023).

The Monday Play: 'When the Wind Blows' (1983). Available at: https://genome.ch.bbc.co.uk/9144b6c80 b9b4c7c97d8da016c46f047 (Accessed: 4 August 2023).

The Nobel Prize in Physics (1903) *NobelPrize.org*. Available at: www.nobelprize.org/prizes/physics/1903/becquerel/facts/ (Accessed: 8 September 2023).

The Notes of Last Intent (2023). Available at: www.youtube.com/watch?v=FuvG53T1gWQ (Accessed: 18 September 2023).

The nuclear phase-out in Germany (2023) *BASE*. Available at: www.base.bund.de/EN/ns/nuclear-phase-out/nuclear-phase-out_node.html (Accessed: 11 September 2023).

The Simpsons (1990). Gracie Films, 20th Century Fox Television, 20th Television Animation.

The Snowman (1982). Snowman Enterprises, Channel 4 Television Corporation, TVC London.

The Sound of Music (1965). Robert Wise Productions, Argyle Enterprises.

The Un-Broadcast: No one is coming to save you (2023). Available at: www.youtube.com/watch?v=cnK_pdEQQJM (Accessed: 11 September 2023).

United Nations Treaty Series (2014), 3007. Available at: https://treaties.un.org/doc/Publication/UNTS/Volume%203007/v3007.pdf (Accessed: 13 April 2023).

vocabulary.com (2023) Megadeath – Definition, meaning & synonyms. Available at: www.vocabulary.com/dictionary/megadeath.

Volta Create (2023). Available at: www.volta-xr.com/ (Accessed: 4 June 2023).

Walker, N. (2023) 'Brexit timeline: Events leading to the UK's exit from the European Union'. Available at: https://commonslibrary.parliament.uk/research-briefings/cbp-7960/.

Wheatley, S., Sovacool, B. and Sornette, D. (2017) 'Of disasters and dragon kings: A statistical analysis of nuclear power incidents and accidents', *Risk Analysis*, 37(1), pp. 99–115. Available at: https://doi.org/10.1111/risa.12587.

When the Wind Blows (1986). Meltdown Productions, British Screen Productions, Film Four International.

When The Wind Blows – Interview with Director Jimmy T. Murakami (2020). Available at: www.youtube.com/watch?v=dXTm6B8nkNA.

When The Wind Blows – Nuclear Snowman (2017). Available at: www.youtube.com/watch?v=0mTJ3keT6qY.

Wiederhorn, J.W. (2022) *39 years ago: Dave Mustaine fired from Metallica, Loudwire*. Available at: https://loudwire.com/dave-mustaine-fired-from-metallica-anniversary/ (Accessed: 7 April 2023).

Winfrey, T. (2022) 'Chernobyl vs. Zaporizhzhia nuclear plants: Why the latter is safer and doesn't spread too much radiation?', *Science Times*. Available at: www.sciencetimes.com/articles/36445/20220304/chernobyl-vs-zaporizhzhia-nuclear-plants-why-latter-safer-doesnt-spread.htm (Accessed: 10 January 2024).

Zagorski-Thomas, S. (2022) *Practical musicology*. London: Bloomsbury Academic.

18

COMPUTER-ASSISTED MUSIC AS MEANS OF MULTIDIMENSIONAL PERFORMANCE AND CREATION

A Post Approach to "Singularity Study 3"

Henrique Portovedo

18.1 Introduction

The intertwining of computation with artistic environments leads to a state of permanent articulation and supports the development of artistic creation. We are immersed in computation, living in a post-humanistic and post-digital world, in which it becomes fundamental to artistic practice, to artworks, and the aesthetic experience. The integration of digital technology and mechanical instruments would not only deconstruct this distinction between electronic and instrumental music, enormously amplifying the scope of extended and augmented techniques, but would also question the traditional understanding of composer, performer and programmer, and their interrelationships. In cases in which the composer and programmer are distinct members of the creative process, insofar as coding affects the compositional options, authorship is deconstructed inasmuch as it does not result from a single mind.

Although instrumental music is by definition technologically mediated, digitalisation has fundamentally changed music production, transmission, and reception, in ways perhaps not fully foreseeable at its origins. The work "SCAMP Singularity 3" is presented here as belonging to a group of works, *Slippery Singularity Studies*, composed from two algorithmic computer systems for electronics and multiple saxophones, namely SCAMP and Slippery Chicken. The first pieces were developed over the specialised algorithmic composition software named Slippery Chicken developed by Michael Edwards, written in and functioning on the principles of the Common Lisp Object System (CLOS), the Common Lisp facility for object-oriented programming. "SCAMP Singularity 3" for alto saxophone and four speakers was developed over SCAMP framework in Python, created by Marc Evanstein, designated to act as a hub connecting the composer-programmer-performer to different resources for playback and notation. The structure and spatialisation of the piece is based on data sonification through the reading of CSV files for data wrangling.

18.2 Post-Humanism and Post-Digital Approach

The concept of post-humanism has various definitions. On the one hand, it is used to designate currents of thought that aspire to surpass humanism, in the sense of the ideas and images of the classical Renaissance. The aim is to update these conceptions for the 21st century, often implying an overcoming of the limitations of human intelligence. Another use of the term post-humanism

DOI: 10.4324/9781003396710-18

is associated with the destiny of transhumanism, the overcoming of intellectual and physical limitations through the technological control of biological evolution itself (Santos, 2020). A physicalist existential state emerges in which the natural transcendence of humanity is sought. Post-humanism is a concept that originated in the fields of science fiction, futurology, contemporary art and philosophy. Post-humanism and technology have evolved together, since the latter is now considered a means of access to knowledge. Social networks, smartphones, etc. are tools that facilitate our access to knowledge as well as to particular sound universes, sonification processes and musical consumption. As post-humanism takes shape in society, hypotheses about the emergence of a new human prototype open up a period of reflection on the promises of technology.

Today, the contemporary landscape, both sonic and visual, is characterised by an omnipresent digital world. Trying to discern between the digital and non-digital domains becomes an intricate task that results in an anticipated ambiguity (Berry, 2014). At the end of the 20th century, Negroponte (1943-) proclaimed the end of the digital revolution, stating that technology no longer had a disruptive nature (Negroponte, 1998). In the 21st century, this same technology, as it had been characterised, became a common phenomenon, having an impact on individuals at various levels (Cascone, 2000). In terms of music, the 1990s witnessed the emergence of a new movement based on digital practices, fuelled predominantly by "self-taught composers" who expressed their disenchantment with the pursuit of technological "perfection", as if their former fascination had disappeared (Cascone, 2000). Thus, the term "post-digital" is characterised as an aesthetic that emerges from the "failure" of digital technologies, encompassing elements such as glitches, bugs, distortions and aliasing, among others (Cascone, 2000).

The post-digital perspective is based on an analogue world completely permeated by the digital. The use of the prefix "post" before the term "digital" does not refer to the end of digital, but to the moment when the digitalisation of cultural reality is more fully established. In the post-digital point of view, technology is not seen as a mere tool, but as a cultural implication. Post-digital evolves into a concept capable of encompassing diverse connotations, proposed as a term that challenges the surge of the digital revolution (Andrews, 2002), but which also denotes the persistence of a specific trajectory. In this way, the post-digital does not represent the cessation of the digital, but rather its continuation (Cramer, 2014), and encompasses a range of variations of computational intensity within it (Berry, 2014). There is currently a strong focus on the advancement of the digital domain, encompassing both software and hardware development, since this is an indispensable prerequisite in contemporary times, intertwining the digital with the non-digital and the physical domain, encompassing elements that belong to both domains (Ferreira & Ribas, 2020). Consequently, this hybridisation has supported the emergence of innovative performative where the central concern does not lie in openly distinguishing between digital and analogue, but rather in scrutinising the omnipresent impact of digital media on our daily lives, revealing its profound socio-cultural effects (Ferreira & Ribas, 2020). As a result, the post-digital artistic creations highlight the exploration of digital technologies and emphasise the cyber-physical interactions between people, digital systems and the material world as the backdrop against which these creations come to life (Ferguson & Brown, 2016).

18.3 The Analysis Of Sound-Based Music

Starting from the concept of *Soundscape* (Schafer, 1993) in the 1970s, a conscious relationship began about the aural identity of a community. This awakened awareness has made it possible to characterise the world and its various realities from the sound elements that make it up, as a new philosophy of knowledge abstracted from the purely musical formulations that had been in force until then. It was only in the 20th century that the purely musical perspective of sound phenomena began to consider its intrinsic physical and physiological properties as relevant in its research (Olson, 1967). Contemporary music itself, that which in its name is considered "music of invention and research" aims to change the paradigm of organisation of its constituent, formal and morphological elements, coming to be called "sound-based" as opposed to "note-based" music:

> Composed interactions are audible experiences as a music of sound (timbre composition) rather than a music of notes, especially when instrumentalists are involved. Music has moved on from musical notes to the timbral composition of sonic spectra, becoming a sonic art that transcends and collapses the traditional dichotomy between sound material and musical form, allowing timbre to be truly experienced as a construction of forms.

> *(Portovedo, 2021)*

In terms of analysing electroacoustic music, we can answer how the listener's listening experience becomes meaningful. Amid the new timbres and spatial soundscapes, the raw electronics and the conflicting juxtapositions, the electroacoustic listener is instantly making sense of auditory sensations and experiencing meaning. Of course, meaning varies greatly depending on the listener. Meaning is a product of the individual listener's mental process: whatever meaning the listener makes of the auditory experience, that is its meaning for the listener. The analyst must bear in mind that they should not attribute meaning to the physical acoustic attributes of a work or to graphic representations based on analysing signals in which the listening experience is not captured. For analysis to be truly credible, it must be conceived fundamentally in terms of human perception and cognition.

Electroacoustic composers have been able to create and use sounds that extend beyond the physical constraints of traditional musical instruments, with a degree of plasticity that allows the modelling of the sound structure itself to play a major role in the musical form (for example, through the manipulation of partials, transients, spectral envelope and micro-sonic levels of sound construction). Since sound recording dispenses with the need to have the physical origins of a sound (such as a factory or a locomotive) present in the composition or final performance space, electroacoustic resources have facilitated a new relationship between intrinsic and extrinsic meaning in music (Smalley, 1996). All of this is encompassed in what Smalley called acoustic space-form:

> The form-space in acousmatic music is an aesthetically created "environment" that structures transmodal perceptual contingencies through links between sources and spectromorphological relationships. It also integrates specific attributes of musical culture and tradition (such as pitch and rhythm, for example). Acoustic space-form inhabits domains somewhere between lived and represented space and the spaces provided by spectromorphological contemplation – by the perceived and imagined configurations of spectral and perspectival space.

> *(Smalley, 2007)*

The idea of the moment in electroacoustic music is a powerful tool, as it unites ways of thinking about sound objects and a raison d'être for how they are used in time. The idea of sustaining the form of the moment through changing static arrangements can provide the platform for active listening "inside" sound, while, as Jonathan Kramer (1978) has pointed out, it is also a vehicle for profound forms of discontinuity. In his words: "The unexpected is more striking, more significant than the expected because it contains more information." John Dack (1999) observed that "Stockhausen's adoption of the moment form does not have to rule out objectively perceptible processes; they simply refuse to participate in an overall directed narrative curve", pointing to the fact that, as an organising principle, it has the potential to expose a sense of the arrangement of structural components as individual entities and inviting the listener to engage in a questioning and creative "formative" mode of listening. The way we perceive a particular moment depends on distinguishing and remembering previous and surrounding events, or whether there is a recognisable pattern in the morphology of chance that can give us a sense of syntactic coherence. The spirit of the moment too has particular potential in acousmatic music, where individual sound gestures or textual identities can be presented in such a way as to direct attention inwards to the idiosyncratic, fleeting and autonomous spectromorphological qualities of a sound.

Trevor Wishart's (1994) genre of "morphic" form extends from electroacoustic music's potential to develop materials through processes of continuous transformation of sonic identity and timbre. The tools of sound transformation found in electroacoustic music can encourage this kind of approach, since the notions of sonic transfiguration and metamorphosis can take on a new meaning, centred on causality (in the sense of sound as an aural product of interactions between a sound body and some input). This evokes the idea of continuum as a unifying psychological construct (for example, in terms of timbre and spectral types), which, if established, can be understood as starting from a certain state or identity and moving towards another. This has the advantage for the conception of form in that a transformational sense of direction, with possible goal states, can be imagined or anticipated before an arrival state is actually known – at least from a traditional perspective, an important ingredient in musical expectation and the creation and resolution of tension. The ideas of the moment and the morphic perhaps together have a new meaning when it comes to form in electroacoustic music – with some interesting consequences.

When designing digital technological works, many questions are also raised about the sound phenomenon, perhaps the biggest of which is related to how the artificial relationships that will control sound production can be constructed and implemented. This separation between sound production and its control has yet to be defined. The use of artificial intelligence in music is an example of this, as it places it in the dual role of composition and performance agent. For d'Escriván, this is a new situation that challenges the traditional concept of the musical instrument, since this definition does not include the instrument as an autonomous participant in the act of musical production (d'Escriván, 2006).

> New technologies mean that we can start with a blank sheet of paper when designing musical instruments and, potentially, the music played through them. This gives us great freedom, but also great challenges. This stems from the fact that, perhaps uniquely in the history of music performance, we are able to separate the production of sound from the means used to control it, (. . .) The state of technological development for much of the 20th century meant that, whatever we might have thought in relation to this debate, the reality was that the art of the performer could make little and, in many cases, no contribution to the realisation of a piece. The technology simply wasn't good enough, (. . .) In this case, a concept of "performance" in real time was simply out of the question in any realistic sense.
>
> *(Miranda & Wanderley, 2006).*

18.4   Active Performance

Performance is distributed, social, contingent and embodied (Small, 1998; Impett, 2001). When playing an instrument there is a distribution of certain requirements and more or less determined sound inclinations (Green, 2014), with the sense of sharing between musicians with the audience as the ultimate goal, in a given physical space, or today also virtually possible. Nowadays, this phenomenon of sharing takes place mainly through virtual means. The physical and acoustic properties of the environment, both of our instrument and of our bodies and/or those of the listeners, are therefore included in the instrument itself, in a very broad sense. All these factors are carefully differentiated in the multiple complexities inherent in performance and which are largely assessed unconsciously by more experienced performers.

With advances in processing speed and the availability of affordable hardware and software from the 1990s onwards, sophisticated musical interaction is no longer restricted exclusively to large institutions (Roads, 1996); works that a few years ago could only be performed in a studio environment can now be included in a concert repertoire. As far as works based on interactive systems are concerned, the quality of the performance is also assessed through the management decisions of these interactive processes, with the performer not only being involved in this management, but also possibly contributing to the dilution of the typical sharp distinction between composer-performer-programmer (Impett, 2001). Karnatz states:

> [I]nteractive technology has become considerably more user-friendly and enabled performers with limited knowledge of the original programming languages access to the process of interaction. The performer can now play the role of composer and make compositional decisions as to the form and content of an interactive work.
>
> *(Karnatz, 2005)*

Throughout my career, responding to different performance challenges, I've had the opportunity to develop know-how in relation to compositional techniques related to mixed music, and therefore to evolve performatively in line with the performance demands associated with this aesthetic. The addition of live components such as foot pedals and other controllers of electroacoustic parameters, mainly aimed at transforming the signal in real time, either on stage or by integrating it into the instrument itself, presents a dilemma, suggesting a hybrid level of performance skill. Interpreters generally don't have advanced knowledge of digital audio signals, acoustics or programming interactive systems, or how to integrate them into performance practice.

Learning an instrument, from beginner to virtuoso level, in the curricula of music conservatories, presupposes an incorporation of the instrument essentially based on acoustic music and therefore focused on that same performance aesthetic. The performance of mixed music and the aesthetic understanding of new music impose different performance challenges and practices. Starting from the definition of performance in contemporary music as a multidimensional action, it is essential to observe how the performative activity of an instrumentalist evolves in the face of current aesthetic challenges. The role of the contemporary instrumentalist is reviewed according to the principles of Active Interpretation (Portovedo, 2020).

18.5 Multidimensionality As Performation

My current musical practice covers various areas of instrumental music and is situated in various aesthetic currents, simultaneously taking on different roles, which can range from interpretation to improvisation, or even programming. The evolution and increase in contemporary thinking regarding interdisciplinarity in the arts, both in creation and artistic research, has increasingly encouraged performers, composers and sound artists to favour the involvement of multiple actors in the composition process, via collaboration (Pestanova, 2008; Nicolls, 2010; Roche, 2011) or improvisation.

One of the requirements for the performance of new music can even be the development of a multi-instrumental practice, something that is often common in jazz, although here with different contours. Re-learning instrumental performance happens through constant performance challenges. This re-learning questions the specialisation of previously acquired instrumental technique, making it evolve while re-evaluating the concept of instrumentality. This process makes it possible to acquire new skills and foster creation through new tools of expression:

> With the advent of the computer, anything that exists can be turned into a musical instrument because the burden of "instrumentality" can be given to the microprocessor. This is to say that as well as how the sound is produced, the gestures that will produce it can be designed arbitrarily by the composer/instrument designer. This designation can redefine dramatically how effort is to be expanded in the production of sound, in fact, it can obviate it altogether; the performer becomes a musical teleoperator.
>
> *(d'Escriván, 2006)*

The comparative representation analysed here is constituted through the intersection of two main areas and, as a reference framework, it is fundamental to identify two opposing tendencies with regard to performance practice, namely: Performance Model and Performation Model, each of which function as gravitational forces with different characteristics and qualities (cf. Figure 18.1).

The Performance Model is associated with more traditional performance practice and is characterised by: the manipulation of an acoustic instrument, with acoustic sound as its output; the visual presence of the performer and the identification of the instrument's focus of sound emission; interpretation based on a conventional notational system; the direct relationship between gesture and sound production from the instrument and the soloist.

The Performation Model is directly related to Active Practice and performative augmentation in an agglutination of the words Performance and Action. The term has been used extensively by Samuel Bianchini in the context of the enjoyment of works of art through the manipulation of an Apparatus (Bianchini & Verhagen, 2016). In this context, the adaptation of the concept to the performative realisation of a musical work is understood. This performance model is, as mentioned above, associated with the control of parameters through the augmentation

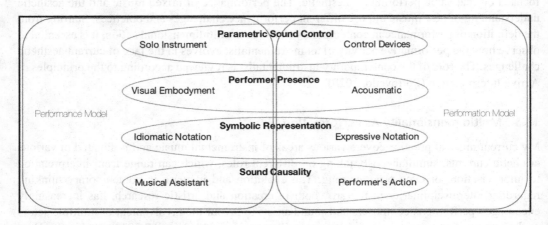

FIGURE 18.1 Performative models.

of performance, expanding the role of the performer and their relationship with materials of increased expressiveness. Some examples can be considered: the use of improvisation; interpretation of unconventional expressive notation; exploration of non-idiomatic elements; co-creative idiosyncrasies between performer and composer. This approach and technique allow for the creation of works that exist in a dimension outside the traditional sound concerns of idiomatic writing (Heaton, 2012). The comparative relationship between the two models and the criteria for evaluating the performance of the works as case studies are defined through the categories: Parametric Sound Control; Performer Presence; Symbolic Interpretation; Sound Causality.

In this analysis model, each category is divided into two variants. When trying to classify a work using this system, the evaluation of one variant does not exclude the existence of another, which in this sense makes it a model of complementary proportions. The quantification of each variant is carried out with a view to attributing a correlation of qualities to the different musical parameters, which can be visualised through a grey bar that fills the graph. To this extent, each work is analysed by both the composer and the performer, assessing the proportions of the different compositional and performative materials present in the work.

Parametric Sound Control is divided between the execution of the solo instrument in a conventional way and how its output contributes to the electronic realisation. In the opposite variant, we refer to the control of external devices, instrumental sensors or the incorporation of gesture and movement sensors necessary for the sound result. *Performance Presence* can even be understood through different prisms: on the one hand, traditional performance practice is based on visual embodiment, playing an instrument implies significant embodied knowledge, "When playing an instrument embodied knowledge plays a significant role as it would be impossible to apply the necessary fine motor skills if all motions were consciously reflected" (Ciciliani & Mojzysz, 2014). It is therefore considered that embodiment has an intimate relationship with sound gesture; on the other hand, Acousmatic performance is related to the performer's non-corporeal presence on stage and the non-direct identification of the focus of sound emission. Electronic music, even if it is spatialised in stereo, often has variations in the focus of the projection of a sound source, for example through panning effects, which allow the sound to travel through space. *Symbolic Representation* evaluates the musical notation of the work for its interpretation. Contemporary sound language presents a panoply of representations when notated; while some works are notated in a conventional way, others are characterised by improvisatory spaces or sections, sometimes free and/or sometimes conditioned. Notational systems have developed in an attempt to favour more direct communication, favouring expressiveness in analogy to the soundscape present in cacophonous everyday life. Experimentation, essential to the current compositional method, not only favours the relationship between performer and composer in a symbiosis of sharing but also explores ways of constantly searching for voices and timbres in the instrumental universe. We can even consider that today a classical musical work is also a scientific work of research and experimentation. The last category, *Sound Causality*, explores the levels of control of the sound parameters, basically assessing whether the sound control and management of the elements that make up the work are entirely carried out by the performer in triggering or similar processes, or whether on the other hand, as in the early days of electroacoustic performance, this management is carried out by another person or system, other than the solo performer.

This framework can be applied to any mixed work, making it possible to carry out a qualitative assessment aimed at developing methods of composition and performance, above all involving the technological character.

18.6 "Singularity Study 3"

"SCAMP Singularity 3A" for Baritone Saxophone and Ambisonics was developed over SCAMP framework in Python, created by Marc Evanstein, designated to act as a hub connecting the composer-programmer-performer to different resources for playback and notation. The structure and spatialisation of the piece is based on data sonification through the reading of CSV files (cf. Figure 18.1). SCAMP provides functionality for managing the flow of musical time, playing notes via SoundFonts or MIDI/OSC messages, enabling quantisation and exporting the result to musical notation in the form of MusicXML or LilyPond.

In this particular case, the Phyton package was used in three dimensions: structuring the macro-form of the work; sonifying the information collected through the CSV files in order to generate the soloist score (cf. Figure 18.2); to create musical gestures based on the different samples developed over the Polyphone software as SoundFonts (sf2 files). In relation to these samples, a SoundFont entitled Frenzy Grains was created, made up of three samples and presents, labelled as: Frenzy Loop, Grains and Grains Low. These samples were developed over the recording of multiphonic permutations within the alto saxophone, as part of the author's research into possible permutations

```python
from scamp import *
from scamp_extensions.pitch import Scale
from scamp_extensions.utilities import TimeVaryingParameter
import random

# read in the data
with open("Data/multiTimeline.csv") as multiTimelineFile:
    lines = multiTimelineFile.read().replace("<1", "0").split("\n")

split_lines = [l.split(",") for l in lines[3:-1]]

drogas = [float(x[1])/100 for x in split_lines]
contraceptivos = [float(x[2])/100 for x in split_lines]

def drogas_part():
    for drogas_value in drogas:
        synth.play_note(int(45 + drogas_value * 20), 0.1 + 0.5 * drogas_value, random.uniform (0.1, 2))

def contraceptivos_part():
    for contraceptivos_value in contraceptivos:
        saxophone.play_note(int(40 + contraceptivos_value * 30), 0.3 + 0.7 * contraceptivos_value, random.uniform (0.1, 2))

s = Session()
s.tempo = 120

# s.fast_forward_in_beats(500)

synth = s.new_part("synth")
saxophone = s.new_part("baritone sax")
voice = s.new_part("voice")

scale = Scale.octatonic(40)

s.start_transcribing (saxophone)

# transcribing different parts
# perf1 = s.start_transcribing (synth)
# perf2 = s.start_transcribing (saxophone)
# perf3 = s.start_transcribing (voice)

INTRO_DURATION = 40

mid_degree = TimeVaryingParameter([20, -10, 0, -10, 20], [INTRO_DURATION/4] * 4)
spacing = TimeVaryingParameter([2, 8, 20, 2], [INTRO_DURATION/3] * 3)
max_wait = TimeVaryingParameter([5, 3, 0.5, 8], [INTRO_DURATION/3] * 3)
average_length = TimeVaryingParameter([8, 5, 1, 8], [INTRO_DURATION/3] * 3)

while s.time() < INTRO_DURATION:
    wait(random.uniform(0, max_wait()))
    degree = random.randint(int(mid_degree() - spacing()), int(mid_degree() + spacing()))
    duration = random.uniform(average_length() * 0.5, average_length() * 1.5)
    voice.play_note(scale[degree], [0, 1.0, 0], duration, blocking=False)

fork(drogas_part)
fork(contraceptivos_part)
wait_for_children_to_finish()

Performance = s.stop_transcribing(saxophone)
```

FIGURE 18.2 Sonification code and Ambisonics performative app.

between them for each saxophone. As mentioned above, coding in Phyton allowed the various samples to address musical meaning by constructing gestures based on pitch, duration and intervals.

Regarding the structure of the work, it consists of three sections: a synthetic introduction based on information generated by SCAMP, later exported as MIDI and assigned to a series of VSTs (Virtual Instruments) using a DAW (Digital Audio Workstation); the central section where a contrapuntal discourse takes place between the solo saxophone (resulting from the sonification of the first data table) and a bass line (resulting from the sonification of the second list of data); finally, the ending section is putting together elements from the previous sections, based on algorithmic selection, through the software designed for the performance of the piece. In this work, the sofinication process consists of quantitatively assigning values to the musical attributes of pitch, duration and rhythm.

The performance of the piece is mediated by software developed in Max/MSP (cf. Figure 18.2), which performs algorithmic and generative functions. The composite electronic sound elements are distributed throughout the space using multiple possible configurations of the ambisonics systems. In this example, an octaphonic configuration is used, in which the various sound sources are renewing their positioning each time a new preset is triggered. The soloist element, the sound of the saxophone, is moving around the space in real time in an automated way.

The performance of the work is based on the musical structure generated by programming in Python using SCAMP. Although the solo part has notated elements, the dialogue with synthetic electronics and with generative and automated spatialisation processes gives way to a discourse integrating improvisatory elements. The fact that these synthetic elements are based on saxophone sounds, particularly multiphonics, means that the perception of the listener is manipulated, giving the impression that many of the electronic sounds are generated from the input of the live instrument. In short, we can suggest a model of performative analysis with interesected elements between the Performance Model and the Performation Model with greater weight of the latter (cf. Figure 18.3).

While "Singularity Study 3" constitutes a highly structured, heavily technology-reliant composition, audiences perceive it as something somehow free and diluted, almost improvised by the computer due to all the algorithmic processes involved. The perceived improvisation contributes to the flow of the performance and smoothly the sonification of data into the performative gestures and sound outputs.

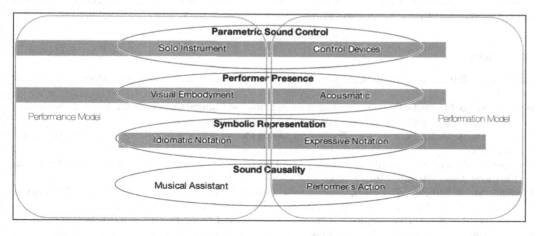

FIGURE 18.3 Performative analysis of "SCAMP Singularity 3".

The sound material of the work gathered from various studio recordings originated an acousmatic sound installation named Lo Incognoscible, based on the book *A Lover's Discourse*, written by R. Barthes and presented at the University of Barcelona. It's possible to access the work at https://youtu.be/W4PB34Nh3q8.

18.7 Conclusion

Technology is moving faster than musical practices and we are taking some snapshots of techniques applied in musical composition and performance, techniques whose materialities will be quickly replaced with new ones, but whose embodied structures continue and become re-implemented in later technical objects as a recycling of skills. Understanding how emerging digital musical technologies trace their concepts, design and functionality to practices in the current cultural epoch will bring to light a study of new-media archaeology, conceptual epistles and performative paradigms, directed, in other words, to the study of how the new technologies of mixed music-making trace their design to the practices of material, symbolic, signal inscription, listening experiences and how practice is transforming and leading to creation.

Understanding how emerging digital music technologies trace their concepts, design and functionality in relation to the practices of the current cultural epoch brings to light an archaeological study of new media, performative paradigms, sound material and symbol inscription practices, as well as techno-humanising creation processes. Do we or do we not see the analogue world with digital eyes, just as we hear mechanical sounds only through digital processes? To what extent do digital media and technologies characterise us in such a way that we can't help but reflect on this aspect in sound and musical art? It is the sum of these phenomena and points of friction in terms of content that motivate me to approach this thematic field. On the one hand, multimedia composition or (even more generally) media art is seen as the most consensual tool, aesthetic or medium for dealing with these themes; on the other hand, it can also be said that these themes are so socially and culturally present, contemporary communication platforms and techniques so imprinted by them, that it is impossible to ignore them in current artistic processes.

Acknowledgements

Henrique Portovedo is a member of the INET-md research group at the University of Aveiro, funded by national funds through the Portuguese Foundation for Science and Technology.

References

Andrews, I. (2002). *Post-Digital Aesthetics and the Return to Modernism*. www.ianandrews.org/texts/post dig.html.

Berry, D. M. (2014). Post–Digital Humanities: Computation and Cultural Critique in the Arts and Humanities. *Educause Review*, 49(3), 22–26. http://dhdebates.gc.cuny.edu/debates/text/20.

Bianchini, S., & Verhagen, E. (2016). *Practicable: From Participation to Interaction in Contemporary Art*. MIT Press.

Cascone, K. (2000). The Aesthetics of Failure: "Post-Digital" Tendencies in Contemporary Computer Music. *Computer Music Journal*, 24(4), 12–18.

Ciciliani, M., & Mojzysz, Z. (2014). *Evaluating a Method for the Analysis of Performance Practices in Electronic Music*. Paper Presented at the International Conference on Live Interfaces.

Cramer, F. (2014). What Is 'Post-digital'? *A Peer-Reviewed Journal About: Post-Digital Research*, 3(1), 10–24.

Dack, J. (1999). *Karlheinz Stockhausen's Kontakte and Narrativity*. https://econtact.ca/2_2/Dack.htm.

d'Escriván, J. (2006). To Sing the Body Electric: Instruments and Effort in the Performance of Electronic Music. *Contemporary Music Review*, 25(1/2 February/April), 183–191.

Green, O. (2014). *NIME, Musicality and Practice-Led Methods*. Paper Presented at the NIME.

Heaton, R. (2012). Contemporary Performance Practice and Tradition. *Music Performance Research*, 5(Spec), 96–104.

Impett, J. (2001). *Interaction, Simulation And Invention: A Model for Interactive Music*. Paper Presented at the Proceedings of ALMMA 2001 Workshop on Artificial Models for Musical Applications.

Karnatz, R. A. (2005). *Interactive Computer Music: A Performer's Guide to Issues Surrounding Kyma with Live Clarinet Input*. (Doctor of Musical Arts). Louisiana State University,

Kramer, J. (1978). Moment Form in the Twentieth Century Music. *The Musical Quarterly*, 64(2), 177–194.

Ferguson, J. R., & Brown, A. R. (2016). Fostering a Post-Digital Avant-Garde: Research-Led Teaching of Music Technology. *Organised Sound*, 21(2), 127–137. https://doi.org/10.1017/S1355771816000054.

Ferreira, P., & Ribas, L. (2020). Post-Digital Aesthetics in Contemporary Audiovisual Art. In M. Verdicchio, M. Carvalhais, L. Ribas, & A. Rangel (Eds.), *XCOAX 2020: Proceedings of the Eight Conference on Computation, Communication, Aesthetics & X*. (pp. 112–124).

Miranda, E., Wanderley, M. (2006). *New Digital Musical Instruments: Control and Interaction beyond the Keyboard*. A-R Editions.

Negroponte, N. (1998, December 1). Beyond Digital. *Wired*, 6(12). Retrieved from https://web.media.mit.edu/~nicholas/Wired/WIRED6-12.html [accessed September 2024]

Nicolls, S. L. (2010). *Interacting with the Piano*. (PhD). Brunel University London, Retrieved from http://bura.brunel.ac.uk/handle/2438/5511.

Olson, H. F. (1967). *Music, Physics and Engineering* (Vol. 1769). Courier Dover Publications.

Pestanova, X. (2008*). Models of Interaction in Works for Piano and Live Electronics*. (Doctor of Music). McGill University.

Portovedo, H. (2020). *Performance Musical Aumentada: Prática Multidimensional enquanto Co-Criação e Hybrid Augmented Saxophone of Gestural Symbiosis*. (PhD). Universidade Católica Portuguesa.

Portovedo, H. (2021). Audible (Art): The Invisible Connections. *JSTA, Journal of Science and Technology of the Arts*, 13(1), 9–20.

Roads, C. (1996). *The Computer Music Tutorial*. MIT Press.

Roche, H. (2011). *Dialogue and Collaboration in the Creation of New Works for Clarinet*. (Doctoral Thesis). University of Huddersfield, Retrieved from http://eprints.hud.ac.uk/id/eprint/17512/.

Santos, A. (2020). *Tecnonatureza, Transumanismo e Pós-Humanidade: o Direito na hiperaceleração biotecnológica*. JusPodivm.

Schafer, R. M. (1993). *The Soundscape: Our Sonic Environment and the Tuning of the World*. Inner Traditions/Bear & Co.

Small, C. (1998). *Musicking: The Meanings of Performing and Listening*. Wesleyan University Press.

Smalley, D. (1996). The Listening Imagination: Listening in the Electroacoustic Era. *Contemporary Music Review*, 13(2), 77–107.

Smalley, D. (2007). Space-Form and the Acousmatic Image. *Organised Sound*, 12(1), 35–58.

Wishart, T. (1994). *Audible Design*. Orpheus the Pantomine.

INDEX

Printed in the United States
by Baker & Taylor Publisher Services